ROLLS-ROYCE
THE CARS & THEIR COMPETITORS
1906–1965

By the same author:
The Lea-Francis Story (Batsford, 1978)

ROLLS-ROYCE

THE CARS & THEIR COMPETITORS
1906–1965

A. B. Price

B. T. Batsford Ltd, London

First published 1986
© A.B. Price 1986

All rights reserved. No part of this publication
may be reproduced in any form or by any means
without permission from the Publishers

ISBN 0 7134 5002 9

Printed in Great Britain by Anchor Brendon Ltd,
Tiptree, Essex
for the publishers,
B T Batsford Ltd,
4 Fitzhardinge Street,
London, W1H 0AH

Contents

Introduction 7
Acknowledgements 9

1 1906–1914: The Silver Ghost 1907–1914 and the Napier 60 1906–1910   11
2 1919–1924: The Post-war Silver Ghost and the Lanchester 40   20
3 1922–1928: The 20 and the Sunbeam 20.9   28
4 1926–1930: The Phantom I and the Sleeve Valve Daimler   45
5 1930–1935: The Phantom II and the 8 Litre Bentley   58
6 1929–1936: The Rolls-Royce 20/25 and the Hispano Suiza H.S. 26 and 30/120   64
7 1934–1939: The Bentley $3\frac{1}{2}$ and $4\frac{1}{4}$ Litre and the Bugatti Type 57   72
8 1936–1939: The Phantom III, the 25/30 and Wraith   86
9 1946–1954: Post-war First Generation   94
10 1955–1965: Post-war Second Generation   116
11 Examples of Roll-Royce Engineering Style   141

Appendix I: Rolls-Royce, Sunbeam and S.S. Cars   149
Appendix II: Notes on some of the engineering personalities   157
Appendix III: The relationship between Rolls-Royce and other manufacturers   171
Appendix IV: Rolls-Royce's surveys of competitors' cars   173
Appendix V: Internal Rolls-Royce correspondence about Bugatti   181
Appendix VI: Correspondence between Dr Lanchester and Percy Martin   185
Index   189

Introduction

The number of books that have been written on the subject of Rolls-Royce Motor Cars is legion. However, a detailed appraisal of their merits, compared with their competitors through the ages, may be of interest.

The problems involved with the repair and restoration of the first two generations of post-war cars are also dealt with briefly, but technical descriptions have been simplified.

Readers will appreciate that the opinions expressed in this book are solely those of the writer, and are gained from being closely involved in the service and repair of Rolls-Royce and Bentley cars for 30 years. He has also been fortunate in being able to examine closely and test all the serious British competitors of the cars from Derby and Crewe, together with a range of high-grade productions from Europe and America. Detailed technical specifications are not included because they have already been given *ad nauseam*; rather, the development of individual components of the chassis have been singled out in order that readers may become familiar with Rolls-Royce engineering practice, particularly in detail work.

The cars of other makes used for comparison are selected as directly competitive. There are many other models which one could consider; they are however, not often readily accessible for study. Other makes which may show up quite favourably have been rejected because they were made in very small numbers, and thus never became serious competitors – the Leyland Eight of 1920 for example.

The book does not cover the third generation of post-war cars which began with the original Silver Shadow in late 1965; these cars are still current and probably not yet of interest to students of motoring history.

Acknowledgements

The author would like to thank the following for their help: Ronald Barker, Esq., The Lord Montagu and the staff of the National Motor Museum, Lt. Col. Barrass, O.B.E. and the staff of the Rolls-Royce Enthusiasts Club, The Henry Royce Foundation, The Lanchester Polytechnic, Coventry, The Birmingham Museum of Science and Industry, H. G. Conway, C.B.E., W. M. Heynes, C.B.E., Dr Noel Tait, Donald Bastow, Alec Harvey-Bailey, Anders Clausagor and the staff of British Leyland Heritage, R. Frost and the officials of the S.T.D. Register, The Earl of Moray, A. L. Swinden, Col. J. L. Cozens, M.B.E., T.D., G. Pilmore-Bedford, Uwe Hucke, The late P. G. J. Hunt, J. M. Fasal, D. E. A. Evans, C. W. P. Hampton, D. L. Taafe.

CHAPTER I

1906-1914
The Silver Ghost 1907-1914 and the Napier 60 1906-1910

The commencement of motor car manufacture by the electrical crane maker Henry Royce in 1904, followed shortly afterwards by his amalgamation with the sales orientated Hon. C. S. Rolls, son of a South Wales coal magnate, has been extensively documented elsewhere. One can describe these rather tentative excursions into the realm of motor car manufacture as an experimental period. Several models were produced and, with one exception, they made for a logical range. Extensions of size was by the addition of cylinders, starting with two and ending with six, the interim models utilising engines of three and four cylinders. It seems strange that Royce found it necessary to bother with a three cylinder car, but his four cylinder production was very successful and won the Isle of Man T.T. Race in 1906.

The six cylinder car was not a success and suffered from severe crankshaft vibrations, resulting in premature breakage of this component, a problem which plagued many early manufacturers of this type of engine. The study of six cylinder crankshaft periodic torsional vibration was in its infancy, and despite the fact that power outputs and engine speeds were very low, crankshafts of the period looked absurdly slender, with no counterbalance weights. Furthermore, they were housed in aluminium crankcases singularly lacking in stiffness, while the cylinders were either cast singly or in pairs and were thus no aid to stiffness either.

The remaining Rolls-Royces of this period bordered on the absurd, consisting of a V8 cylinder car with the engine amidships and restricted to a point where it would not exceed 20 m.p.h on the level, then the legal limit. The outstanding finish and attention to detail which was to characterise the Silver Ghost at a later date is not so apparent in these early cars. It would be fair to say, for example, that they were inferior to the products of Napier, although the Acton firm had already enjoyed several years of experience in the field of motor car production. However, the announcement of the Silver Ghost in 1907 immediately placed Rolls-Royce in the forefront of the motoring world. The former electrical engineer, Henry Royce, had produced a greatly improved machine. The car was conventional in design, but of greater refinement, and this last feature was constantly embellished until 1914. Cars built between 1919 and 1925 show little further development, and do not appear quite as attractive as the pre-war models with their low set radiators and generally lighter appearance. Towards the end of production, the Silver Ghost began to look distinctly outdated; indeed, front wheel brakes were not fitted until the final chassis were being built, although many cars were subsequently modified by the works to incorporate this overdue improvement. The technical features of the Rolls-Royce Silver Ghost have been described in many books, but it appears to have escaped the notice of most that an almost complete re-design of the chassis took place in 1911 and the year after. The original model presumably revealed shortcomings, and it appears that the serious single-minded pursuit of perfection did not begin until the second design was formulated.

The 1907 Silver Ghost retained by Rolls Royce Ltd (in company with an experimental Bentley Mk VI).

Rolls-Royce three cylinder engine, 1905. Note the early form of the Royce-designed carburettor.

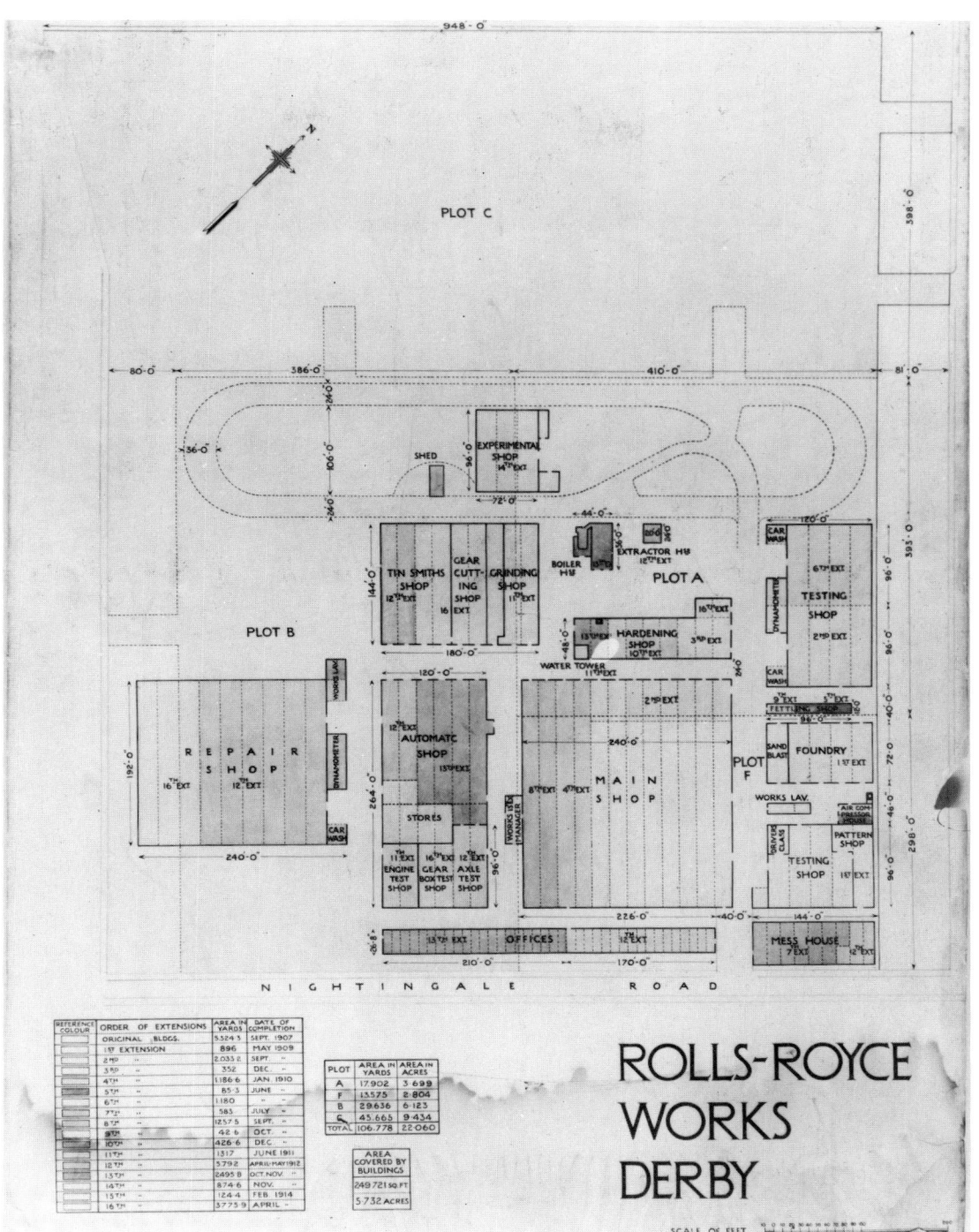

Factory plan of the Rolls-Royce works at Derby showing no fewer than 16 extensions between the opening in September 1907 and April 1914.

The original Derby works.   An early silver Ghost rear Axle.

Slightly later design of Silver Ghost rear axle.

The chassis frame itself was stiffened, while the original crude brake operating layout was abandoned. This had consisted of cables which were forced around tight radii as they wound their way, snake-like, along the operating levers.

The later design revealed conventional cables ending in yokes and clevis pins incorporating neat ferrules through which the split pin passed; a nice design feature, sadly abandoned in 1945.

The rear axle of the early cars appeared delicate and consisted of an aluminium differential centre to which were bolted steel axle tubes. A mere $7 \times \frac{3}{8}$ in. bolts sufficed to hold the two axle halves together. It was, in fact, almost identical to the previous 30 h.p. car. A complete re-design of this component took place, and the opposite extreme was reached with well-shaped steel axle halves now bolted together with no less than $20 \times \frac{3}{8}$ in. diameter bolts, while a torque tube layout was now adopted, and this part was secured to the casing with an amazing $24 \times \frac{3}{8}$ in. diameter bolts; indeed, there is only just sufficient spanner clearance.

The power plant, which was the most impressive part of the original 1907 car, was refined in detail.

The second design aimed to balance the moving parts, and the crankcase was stiffened.

The large-engined Napier from about 1910, built and still used for competition work.

Three entirely different gearboxes were evolved before the First World War, commencing with four speeds, top arranged as an overdrive, followed by a three-speed type, and then the beautifully engineered conventional four-speed type, which was to remain largely unchanged for many years.

The Silver Ghost was in direct competition with the larger offerings of the Napier Company, particularly their 60 h.p. model. This car was designed some eighteen months before the Silver Ghost, and while the chassis of both cars are broadly similar, the engine of the car from Acton appeared to be the product of an earlier generation. It was, in fact, the first really successful six cylinder design, and embodied three blocks of two cylinders against the two-blocks-of-three layout of the Silver Ghost. The Napier was a little larger, at 7,724 c.c. ($4\frac{1}{2} \times 5$ in. bore and stroke), against the Ghost, originally of square $4\frac{1}{2} \times 4\frac{1}{2}$ in. dimensions and 7,036 c.c., but enlarged by 1909 to 7,428 c.c. by dint of increasing the stroke by ¼ in. The overall length of the Napier engine was considerable at some 4 ft 4 in. (3 ft 4 in. between the outer of the four engine bearers). The Derby engine was larger still at 4 ft 8 in. overall (3 ft between engine bearer centres, two bearers only).

The Napier enjoyed a performance greater than that of the Ghost in Standard form, but in order to underline this fact a larger version was evolved which measured 4 ft between crankcase bearer centres and with bore and stroke of 5×6 in. respectively (11,580 c.c.). The racing cars thus fitted offered a prodigious performance which was delivered with a surprising refinement. The Napier induction system appeared tortuous, with the carburettor on the offside nestling between No. 4 and 5 cylinders; a pipe ran through the engine, joining a riser on the nearside which led into the centre of a pipe whose ends let into a further pipe immediately below, which fed into ports in the cylinders.

The crankshaft looked alarmingly frail.

The con rods were provided with four big end

Napier 40/50 chassis assembly, Acton works, July 1920.

Napier 40/50 engine mounted in a large centre lathe. The flywheel housing machining operation is in progress.

bolts, a Napier feature. The appearance of the car was marred for some by the forward mounting of the radiator in order to clear the enormous engine. It was, in fact 12 in. ahead of the axle centre line, although in the racing cars it was moved rearwards by 2 in. This position was unique to the Napier, and did not reappear until 1931–1932 when the Lanchester 18 and Wolseley Hornet adopted the forward radiator, hailed, of course, as a new feature in order to provide more passenger space within a given wheelbase.

A serious attempt to hold down weight was apparent in the Napier chassis; all forgings were well-proportioned and hollowed out in low stress areas. The suspension systems on both cars were originally identical, embodying semi-elliptic springs fore and aft, together with an additional transverse spring at the rear; the roll problem must have been considerable when fitted with limousine coachwork.

The Rolls-Royce concern soon dispensed with this arrangement in favour of the marginally better three-quarter elliptic rear layout, a system adopted by Morris at the other extreme of the market.

For 1912 the definitive Rolls-Royce layout appeared, embodying cantilever rear suspension in the Lanchester mould.

The Napier racing cars also dispensed with the additional transverse spring.

The rear axle evolved by Napier consisted of well proportioned cast steel half casings joined on the axle centre line by a mere $6 \times \frac{3}{8}$ in. bolts identical to those in the early Ghost, while the brake actuation by rods and shafts incorporated a swivelling compensator, a somewhat coarser solution than the elegant, if expensive, little gear differential compensators to be found on the Silver Ghost.

While the Napier design stagnated to a degree, Rolls-Royce embarked upon their refining policy, and had very definitely stolen a lead over the London firm by 1912.

Napier attempted a comeback after the First World War with an overhead camshaft $6\frac{1}{2}$ litre car of great technical interest. Curiously, no great effort appears to have been made to market this car in an aggressive manner; indeed, its sponsors seemed to loose interest, and it faded quietly away in 1924 after a mere 180 cars had been built. The energies of the firm were then concentrated upon their aero engine business.

CHAPTER 2

1919–1924
The Post-War Silver Ghost and the Lanchester 40

The name of Lanchester became pre-eminent as a manufacturer of high grade and individualistic motor cars in the pioneer period prior to the first world war. Dr Frederick Lanchester, undoubtedly one of Britain's most gifted and innovative engineers, was born in October 1868. He undertook important pioneering work in the field of flight as well as automobile engineering, where his main work was concerned with the problem of engine balance. His interests were wide ranging and included such work as rigging up Birmingham Town Hall for a demonstration of radio in 1905. He founded the Lanchester Engine Co. assisted by his brothers George and Frank. George was responsible for control of the works and ultimately took over design from Frederick, his tutor and elder by six years. Frank Lanchester managed the commercial and sales organisation.

The early years were notable for serious financial problems, causing one receivership and a severe cramp on growth. The Pugh family, who had made a fortune in the cycle industry, through the Rudge-Whitworth business, took control of a reconstructed company in 1905. The works were situated in Sparkbrook, Birmingham and survived until the economic blizzard of 1931. The inevitable collapse was followed with absorption of the assets by the Daimler Company. Almost immediate abandonment of existing Lanchester models was the outcome, while the name was retained for use on various Daimler models, generally at the cheaper end of their market. An early example of the questionable habit of badge engineering, the name continued until the middle '50s in this fashion.

George Lanchester took over as Designer and Chief Engineer in 1914, a post he held until 1936, the last five years under the aegis of the Daimler Company. He left to take a position with the Alvis concern and was responsible for their 12/70 model.

Towards the end of the First World War, work commenced at Sparkbrook on a new 6 litre luxury car with dimensions of 4 in. bore and 5 in. stroke and an R.A.C. rating of 38 h.p. The new car was to be known as the 40. The Lanchester Company, in 1919, held accumulated profits of £180,000 as a result of lucrative War Office contracts. No doubt this relative affluence encouraged the directors to challenge Rolls-Royce head-on. Dr Frederick Lanchester, although not directly consulted over policy or design aspects of the 40 strongly condemned this policy. He knew that the firm lacked the technical resources of Rolly-Royce, and might have addded that the marketing expertise of the latter was also vastly superior.

The Lanchester concern remained a comparatively small business throughout its life, and the fact that their large car constituted a serious threat to Rolls-Royce sales is highly creditable to those concerned. The general performance of the 40 was superior to that of the Silver Ghost, while the whole design, like all Lanchester models, bristled with interesting features. The workmanship was superb and the car represented a high form of engineering artistry. It is surprising that C. S. Rolls did not seek to form a liaison with the Lanchester brothers when casting about for a British car concession. Perhaps

A Lanchester 40, circa 1925. A front-wheel-braked example, and almost indistinguishable from the smaller 21. The latter, however, utilised solid disc wheels until 1928.

An example of a later series Silver Ghost with Barker coachwork originally purchased at the 1922 Olympia Motor Show by a member of the Sainsbury family.

Lanchester 40 engine and gear box test bank.

he did, and failed for some unknown reason. It is fascinating to contemplate the magnificent cars that we may have seen if such a partnership had prevailed.

The power unit of the 40 followed First World War aero engine practice, a popular theme adopted by many high grade manufacturers in this period, but brought to a very high pitch of refinement by George Lanchester and his team. The layout of this power plant consisted of two blocks of three cylinders mounted on an aluminium crankcase in the accepted manner of the time. The crankshaft ran in seven bearings with main bearing diameter of $2\frac{1}{4}$ in., although the big ends were larger at $2\frac{1}{2}$ in. The con rods were well shaped and fully machined, while a copper feed pipe for small end lubrication was provided and clipped into position. An overhead camshaft operated inclined valves via the medium of L-shaped rockers, while valve clearance was adjusted by the fitting of suitably sized valve caps.

The usual thought was given to the question of access to valves. The inlet was placed in a detachable cage which could be easily removed from the cylinder head without disturbing the camshaft. The exhaust valve was not caged for reasons of heat transfer, but could be removed by allowing it to drop on to the piston crown, whence it could be

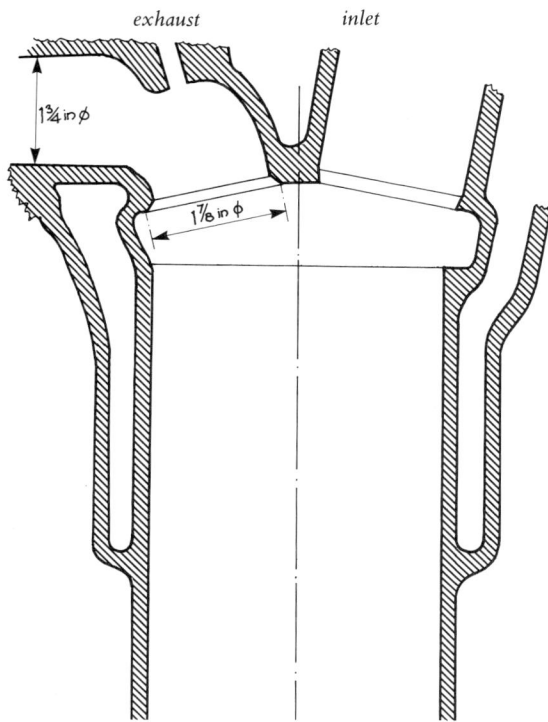

Lanchester 40 cylinder and porting arrangement, the inlet valve cage removed. The valve head diameter (inlet and exhaust) is 1⅞ inches, with a tulip head.

fished out through the inlet cage aperture. The camshaft was driven by a vertical shaft with a typical Lanchester worm drive at both ends. The Lanchester Brothers had an obsession with worm drives, and, of course, used such gearing for all of their back axles dating back to 1900, and in this connection they were compatible with the thinking of the Daimler Company, who had also adopted the Lanchester patent worm. A unique position for some of the accessories was evolved by placing a short shaft parallel with the crankshaft at the nearside rear of the engine, driven by very large spur gears. Upon this shaft were fitted three worm drives in a row for a vertical dynamo, a starter, the latter incorporating a free wheel, and finally an oil pump drive. The engine was a most impressive sight, and completely filled the low bonnet; in fact, the front of the polished aluminium valve cover was almost level with the top of the radiator. The well known Lanchester torsional crank damper was, rather surprisingly, not fitted to the 40 h.p. engine.

The transmission system followed pre-war Lanchester practice, and consisted of a three-speed epicyclic gearbox controlled by a conventional right hand lever, while the system included a conventional single plate clutch which provided direct drive when engaged. The clutch pedal was, in fact, a gear engagement pedal, and a clever lock-out system prevented movement of the gear lever until the pedal was depressed.

The chassis and suspension layout of the Lanchester 40 bore evidence of very careful thought. The chassis frame was much stiffer in torsion than was usual at the time; a very substantial cross member ran behind the gearbox, and contained a 'window' through which passed the front end of the torque tube. A further cross member was positioned ahead of the rear axle which consisted of a tube of no less than 6½ in. diameter. The rear of the chassis side members were united by the petrol tank which was designed as a structural part of the chassis frame, perhaps not a very good idea, but apparently satisfactory in service. The front half of the chassis side rails were boxed in, although the inner wall was perforated with large holes in the interest of weight reduction. The front springs were extremely long at 43 in. between centres, while the axle beam, fully machined of H section, and almost straight, was bolted on top of the springs. Rear springs were of cantilever pattern with free ends which passed between rollers, spring loaded against side float; a very nice piece of detail work. The hand controls at the top of the steering column were neatly recessed into the steering wheel, and the whole car gives evidence of fine craftsmanship and tidy thinking.

The Lanchester 40 was certainly a fascinating car, and was possessed of almost steam engine torque, and uncanny smoothness at slow speeds, although the engine appeared to have one rough period at higher revolutions. The writer, has, however, only experience of one car. The unorthodox transmission system must have been a tremendous boon in the early twenties, and it is impossible to make a bad gear change; one can, in fact, select reverse while travelling forward at 40 or 50 m.p.h. with complete impunity. The system was entirely satisfactory in service, and was years ahead of any of its competitors. The standard of riding comfort was also

Lanchester 40 chassis assembly bay, Armourer Mills, Sparkbrook, Birmingham, 1920.

outstanding, and definitely superior to the Silver Ghost, a facet that enabled one to take advantage of the superior performance of the car.

The valve gear was definitely audible, although within acceptable limits. The radiator was raked backwards at about 3°, an unusual feature at the time, while the quaint hallmark of Lanchesters, the window through which one could see the water level without removing the pyramid shaped cap, was in evidence. This feature might have offended Rolls-Royce, for it revealed just how rusty cooling water became after a very short time, and this murky sight was in full view at the point where others would have fitted a badge. The whole car was, however, distinctive, indeed imposing in appearance.

Production totalled approximately four hundred and continued until 1929, although only a handful were completed after 1926.

The principal British competitor for the Rolls-Royce 20 was also a Lanchester, designed in 1922 as a 20 h.p. car with the purpose of carrying four people in comfort. The layout resembled a scaled down 40 but the Directors insisted on a normal four speed sliding gear with right hand change for the transmission; they were of the opinion that sales were lost due to the unconventionality of Lanchester features. The first engine suffered from a bad period and the slender crankshaft of 1½ in. diameter for both mains and big ends was abandoned; a larger one of 1$\frac{11}{16}$ in. diameter was substituted which would still fit the original crankcase. The bore was also enlarged and the car was launched as a 21 (74.5 × 114 mm.) bore and stroke. The reason for the increase in size was a demand for full six-seater bodies and Lanchester felt bound to compete with the new Daimler and Rolls-Royce 20 h.p. cars, both of which were often burdened with absurdly large bodies. The power/weight ratio of the Lanchester was not considered really satisfactory; Hamilton Barnsley, the Managing Director of Lanchester, in a letter to Dr F. W. Lanchester states, 'We don't seem to have quite rung the bell with the 21'.

The overhead camshaft valve gear differed in detail from the 40 while a monoblock was adopted. Four wheel brakes on the Perrot system were standardised, embodying massive 14½ in. diameter drums. In order to make some distinction with the expensive, larger model, the 21 was fitted with rather inelegant solid disc wheels, which were practical, though not ideal for brake cooling. The usual Lanchester quality was maintained, indeed enhanced in one or two particulars. The layout of the electrical equipment, wiring conduits, fuse boxes and the like were the subject of careful attention and the individual fittings were quite exquisite in their execution.

The Lanchester firm possessed their own coach-building shops and good engineering also entered into this department, a state of affairs unusual at the time. An innovation concerned the adoption of aluminium castings for floor sections, scuttles and door pillars. Daimler also followed this method of construction, probably through Lanchester influence. The bore was enlarged to 78.7 mm. for the 1926 season and a useful gain in torque resulted, although the maximum power was a modest 57

Lanchester 21 power unit. The Lanchester external friction damper is in evidence and the firm's inclination for unusual accessory drives can be seen in the right-angled position of dynamo and magneto. The cylinder block appears to be held down with a minimum of studs.

A 1925 21 h.p. sports model Lanchester. The company commissioned Arthur Mulliner of Northampton to build this style.

b.h.p. at 2,800 r.p.m. The car would now exceed 60 m.p.h. when carrying the heaviest body. A most attractive boat-decked sports model was also offered, on a short chassis and fitted with a higher axle ratio. The 1929 season saw the adoption of the Rudge-Whitworth knock-off wheels for all types and the various body styles became almost sporty in appearance.

The final design from the Sparkbrook firm was in the form of a Straight 8 which was introduced, as were most Straight 8 engines, in late 1928. Lanchester thus slavishly joined the short lived craze for this engine configuration. The new car consisted of a 21 with two further cylinders added, and thus a capacity of 4,440 c.c. Smart fabric sports saloon coachwork of their own design was generally fitted and the traditional radiator was given thermostatically operated vertical slats. This sophisticated car was, however, expensive at £1,775. The car received rave notices in the press with claims of a genuine 80 m.p.h. together with total smoothness throughout the speed range. Nevertheless, only 125 were built, the last few chassis being assembled at Coventry after the Daimler absorption.

Approximately five hundred 21–23 h.p. six cylinder cars were delivered, just over one-sixth of the production of the Rolls-Royce 20, quite a good result taking into account the modest size of the Lanchester business. If Lanchester had been able to command the worldwide sales and service back-up which the Derby company enjoyed, the title of Best Car in the World would probably have been lost to

Birmingham during this period. The failure of the Lanchester Company took place a few months before that of the original Bentley concern. We know that Napier made a concerted effort to purchase the Bentley business, and went so far as to offer approximately £120,000 for the assets, just losing to Rolls-Royce. A trifling £38,000 would have secured the Lanchester business instead!

Lanchester Production and Financial Results

	1926	*1927*	*1928*	*1929*	*1930*	*1931*
No. of cars sold	271	223	198	215	176	45
Profit/Loss	£8,000 Profit	£3,000 loss	£8,000 loss	£8,000 loss	£10,000 loss	(Unknown – consolidated with Daimler accounts)

Lanchester 21. A late example from 1929–1930 attempts to promote a more sporting image, as can be seen with the adoption of Rudge wire wheels and the slightly strange close-coupled fabric sports saloon coachwork.

CHAPTER 3

1922–1928
The 20 and the Sunbeam 20.9

It is interesting to compare the small Rolls-Royce with the effort of another Midland manufacturer, the Sunbeam Motor Car Co. Ltd, of Wolverhampton. This business, which represented the largest arm of the Sunbeam-Talbot-Darracq combine, exhibited a strong continental flavour by virtue of the influence of the head of engineering, Louis Coatalen. The firm, much expanded during the 1914–1918 war, enjoyed very well equipped facilities, including their own foundry, good laboratory facilities and, like Lanchester, a coachbuilding department.

1926 20 Barker coupé demonstration car.

This business reached its zenith in the '20s, producing a large number of perhaps too wide a range of expensive and beautifully finished cars.

Sunbeam cars always possessed a distinctly good performance delivered in a highly refined manner; their popularity was such that around sixty cars a week were delivered during peak periods in 1927–1928.

A six cylinder car rated at 20.9 h.p. with a bore and stroke of 75 × 110 mm. (2,916 c.c.) was prepared during 1926 and entered full production for the 1927 season. There had been previous Sunbeam cars of similar general design and size, but this particular type was the most numerous and con-

Early Rolls-Royce 20 h.p. chassis in the test department at Derby. Rear axles were presumably unsatisfactory at this period judging by their absence from certain cars.

tinued through 1930 with but few modifications, the most obvious of which concerned the introduction of a V-fronted radiator for 1927 similar to the 3 litre sports car, and first seen on the 1924 2 litre G.P. racing cars. The 20.9 engine was revived in 1933 for installation in a model known as the Speed 20, a most attractive car which was listed later as a Speed 21 until the inevitable bankruptcy in July 1935.

The Rolls-Royce 20 was introduced in 1922 but was considerably revised and improved during 1925, when a right hand change four-speed gearbox replaced the previous Americanised three-speed centre change pattern, while the car was given four wheel brakes at the same time, incorporating the well known friction servo driven from the tail of the gearbox. This car continued into 1929 and was comparable with the Sunbeam 20 during this period, although a little larger in capacity and much higher in price. A 1928 Sunbeam 20 complete with factory built Weymann pattern fabric saloon was priced at £825, while a Rolls-Royce 20 with similar coachwork constructed by, for example, H. J. Mulliner, was listed at around twice this figure.

The power unit of the Sunbeam was robust, differing from the Rolls-Royce in many respects, but with six cylinders and pushrod overhead valves.

The repair department at Derby in the early '20s.

The Wolverhampton firm utilised a common crankcase and cylinder block of four main bearings configuration in place of the separate aluminium crankcase containing seven main bearings and detachable cast iron cylinder block of the Derby engine. The additional stiffness of the Sunbeam no doubt rendered additional bearings unnecessary, as well as resulting in a reduction of overall crank length. A damper was not required.

The Sunbeam was not quite as smooth as the Rolls-Royce but the latter was aided by an enormously heavy flywheel while power output was deliberately restricted. The performance of the Rolls-Royce 20 is almost non existent whereas that of the Sunbeam was extremely good and cars with lightweight bodies would certainly achieve 75 m.p.h., with acceleration to match. The appearance of the engine of the latter was also much neater than that of the Rolls-Royce, due to the European influence, although the painted wood bulkhead appeared cheap compared with the well finished cast aluminium one always in evidence on pre-war Rolls-Royce cars.

Both cars possessed excellent four-speed gearboxes and efficient braking systems; the Sunbeam system of rods and cables embodied front brake operation via a universally jointed shaft on the Perrot system, and a vacuum servo was provided.

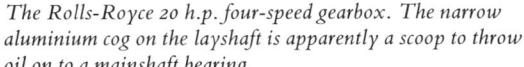

The Rolls-Royce 20 engine.

The Rolls-Royce 20 h.p. four-speed gearbox. The narrow aluminium cog on the layshaft is apparently a scoop to throw oil on to a mainshaft bearing.

Sunbeam 20.9 h.p., UK 6219. The works 1929 Monte Carlo entry driven by Sales Manager Leo Cozens, photographed prior to departure outside the works recreation block. This car survives in totally original condition.

Sunbeam 23.8 h.p., GC 6. The works 1930 Monte Carlo Rally entry embodying new features in the light of experience gained the previous year.

Sunbeam 25 h.p., GC 2. The 1930 Monte Carlo Rally entry of Messrs Pass & Joyce Ltd, London distributors. Note the choice of European lighting equipment on this car.

Sunbeams GC 6 and GC 2 prior to setting out for John o'Groat's, January 1930. Left to right: Leo Cozens (Sales Manager), General Huggins (Director), A.H. Pass and Colonel Iliffe.

Sunbeam 23.8 h.p. dashboard – the 1930 Monte Carlo Rally car. The three-panel windscreen was devised so that the passenger could enjoy some protection in the fog if the driver needed his side open. Note the Rotax switch gear, with individual circuits for all special equipment. The speedometer was printed with an additional dial in kilometres per hour and a battery meter was provided.

Sunbeam 18.2 h.p. touring car of 1931 destined for India (for His Highness the Maharajah of Kedah) photographed outside the Shrewsbury gate of Weston Park, the seat of the Earl of Bradford and often used for sunbeam publicity pictures.

Sunbeam 18.2 h.p. chassis, 1932. Present day restorers should note the comparatively rough paint finish on the chassis frame.

Sunbeam 20.9 h.p. engine and gearbox, nearside. The neat layout is a result of the European influence in the design office.

Sunbeam 23.8 h.p. engine, offside, showing self-cleaning oil filter rotated by depression of the brake pedal. The exhaust manifold appears to be a well-fettled casting.

Sunbeam testers stand behind a Speed 20 at the half-way point on the Wolverhampton-Enville-Bridgnorth test circuit, on a summer's morning in 1933. Note the down-draught Stromberg conversion on the revived 20.9 h.p. engine. A slave valve cover without the fine enamel finish was retained until the completion of testing and assembly.

The same Sunbeam testers near Bridgnorth. A Speed 20 chassis in the foreground, with a 23.8 h.p. behind.

The pedal pressures and effectiveness appeared about equal on both cars, but the mechanical servo of the Rolls-Royce enjoyed the important advantage of continuing to work with a stalled engine. The vacuum servo system of the Sunbeam derived its power from inlet manifold depression and could prove disconcerting, if not downright dangerous, when confronted with a stationary power unit. The Rolls-Royce system also enjoyed a lower level of friction through the linkage employed, although the system was unduly complicated and must have been extremely costly to manufacture.

The rear axle of the Sunbeam utilised a torque tube rendered necessary by the cantilever rear springing, whereas the semi-elliptic springs of the Derby car made possible the use of an open propshaft. This Hotchkiss drive system allowed the torque reaction to take place through the springs, which assisted smooth take up of the drive, although the Sunbeam clutch, similar in construction to the Royce, was very silky in operation.

The roadholding and steering of the Sunbeam was distinctly superior to that of the Rolls-Royce, no doubt one of the benefits of prolonged racing experience. The standard of ride achieved by Wolverhampton also appeared superior.

The great majority of Sunbeam's cars were fitted with in house coachwork, and the standard of work turned out by the Wolverhampton shops was the equal of the great London coachbuilders, with

A 1929 25 h.p. Sunbeam, the property of the Duke of Gloucester. Depicted after a total re-build incorporating a 1934 bonnet and radiator together with new coachwork.

A special Speed 20 fixed head coupé of imposing appearance, in late 1932, outside St Philip's Vicarage, Wolverhampton.

A 1933 Sunbeam Speed 20 showing fitted suitcases and neat tool storage.

A similar car showing the pillarless body, of a type then enjoying a brief vogue.

The driving position in a 1934 23.8 h.p. sports saloon, epitomizing the quintessence of British workmanship.

A late Sunbeam chassis, now with downdraught carburation and modernized ribbon radiator, photographed in the well-scrubbed works – with gas lighting.

1929 20 with the higher radiator and vertical shutters. The Barker coachwork shows evidence of European influence.

beautiful detail work, fine veneering of interior woodwork and attractively fashioned trim. The Sunbeam was a better-looking car than the great majority of Rolls-Royce products, burdened as they were with dreadfully ponderous, formal coachwork, while the ill-fitting line up between radiator and bonnet remained a faintly absurd anachronism. Coachbuilding was not a typical Black Country activity, but the individualistic folk from this district were ever adaptable and no doubt the whole process of building motor cars was punctuated with outbursts of a unique brand of humour. Taking into account the purchase price, undoubted longevity and vastly superior performance of the Sunbeam, one has to admit that it represented much better value than the Rolls-Royce 20.

Star

Wolverhampton was the home of another highly respected maker which mounted a challenge to Sunbeam supremacy during the final years – the Star Engineering Company. Star was an old established firm which passed through the stage of cycle manufacture but was also early in the field of car production by way of copying first Benz and, later, De Dion designs. The firm was owned by the Lisle family and became a major force by 1914, achieving the rank of fifth largest producer in Great Britain.

A concerted attack on the colonial market was comparatively successful and specific models were developed with this in mind. The chief designer, F. A. S. Acres, actually delivered a paper on this subject to the Institute of Automobile Engineers in 1923. Various models between 12 and 20 h.p. were on offer before the company finally settled for a comparatively sophisticated and high grade six cylinder car built in 18 and 20 h.p. sizes. The engine was of push rod overhead valve layout, the crankshaft supported on seven main bearings and provided with a vibration damper. An interesting feature concerned the provision of detachable wet liners while the combined crankcase and cylinder block, if heavy, was an extremely rigid unit. Dural conrods were specified. The Star concern were early users of Fabroil camshaft timing gears and suffered the usual eventual failures. The dynamo and magneto situated on the nearside were driven in tandem

Star 20/60 coupe, 1927.

Star 20/60 limousine, 1928, with Star coachwork, considered to be one of a consignment for King Saud of Saudi Arabia. *This order was obtained by Colonel T. E. Lawrence, who put his Middle East connections to good account.*

Star Comet 18/50, final series, 1931–1932, with Star coachwork.

from the timing gear train, reminiscent of Sunbeam practice. The entire appearance of the engine bore a close resemblance to the standard products of the Moorfield works.

The Star concern embarked upon a serious racing programme in the early years but this policy was abandoned after 1918. They did, however, make a foray into the lorry and coach business and some modest success attended this venture for several years. This proved attractive to Sydney Guy, who bought the business from the Lisle family in July 1928.

The cars continued and styling was improved; by 1929 the Star had become a fine looking car, extremely well equipped and possessed of a highly respectable performance. The quality of workmanship and materials was first class throughout, with neat and well-conceived detail work. If one considers that the Sunbeam 20 represented outstanding value at £875 compared with a Rolls-Royce 20 at £1,475, the Star 18/50, priced at £595, would appear to represent the ultimate in value for money. The result was that the order book remained in an apparently satisfactory state well into the depression years, with production running at a constant 15 cars a week. The fact of the matter was, however, that the cars were sold at a loss, a state of affairs which continued for years, rather similar to the conditions at Lanchester.

The 1931 season saw re-styled cars with thin ribbon radiator shells and oblong instruments, in the Chrysler manner. C. P. Beauvais, one of the first professional stylists, was responsible for the new bodies. F. A. S. Acres had by now left to join the design staff of Vauxhall Motors; his place at Star was taken by Jimmy Pratt, who later joined H. B. Manley, the latter having purchased the Girling brake patents. The only retrograde step by Star concerned the adoption of the erratic if powerful Bendix brake system. Twin-top gearboxes were also specified, then enjoying a short vogue. It was merely a normal gearbox with a high third speed ratio, while these gear teeth were normally of herringbone design in order to achieve silence. The third speed could thus be used as the normal gear for town work. We have heard that these gearboxes were purchased from Humber.

The final 1932 season saw the range extended with a smaller 14 h.p. version (2,100 c.c., 63.5 × 110 mm.) priced at a mere £345 and still embodying all

Special 14 h.p. Star Comet fitted with a McEvoy supercharger and Jensen four-seater sports body. Still extant.

the Star quality features, while the Star Flyer coach engine, of identical layout to the cars but stretched to 3,620 c.c. (80 × 120 mm.) was also offered as a 24 h.p. version. This model was good for 80 m.p.h., but only eight were built. None appears to have survived. The wide ranging price differential between the smaller 14 h.p. through 18 h.p. and 20 h.p. up to 24 h.p. is amazing. The cars were virtually identical in specification, with similar coachwork and accessories. It clearly cost no more to manufacture the large car when the only difference concerned cylinder bore and crankshaft stroke. At £345, the 14 h.p. car would result in a substantial loss for the makers and was presumably a desperate measure to clear a surplus stock of bodies and chassis. At £695 the 24 h.p. car would clearly be highly profitable, but the unfortunate makers were baulked here as well because buyers for this car could not be found. Production ceased with the appointment of a receiver in April 1932.

The cost of building motor car bodies in the vintage years is nowadays of merely academic interest, but students of the economics of the motor industry may find interest in the table below:

Cost of Body Production Make an Interesting Study

Body	Specification	Builder	Customer	Price	Year
Open tourer	Ash body frame ready for panelling	Edward Evens, Dudley	Bean	£2.50	1925
Minx saloon	Fully-trimmed, glazed and painted	Pressed Steel Co.	Humber-Hillman	£25.00	1933
2–4 seater tourer	Complete with hood and side screens, leather upholstery, mounted on chassis	New Avon Body Co.	Lea-Francis	£55.00	1931
Fixed-head coupé	Fabric covered, (under Weymann patents), mounted on chassis, hide upholstery	Cross & Ellis	Lea-Francis	£95.00	1930
4-door Francis saloon	Fabric covered, hide upholstery	Cross & Ellis	Lea-Francis	£73-17-0d.	1930
Cabriolet de-Ville	Cloth rear, hide front, division, occasional seats. Mounted on Rolls-Royce 20 chassis	Barker	Private	£755	1925

CHAPTER 4

1926–1930
Phantom I and the Sleeve Valve Daimler

The successor to the Ghost eventually appeared in 1925; the chassis remained unchanged, but a new pushrod operated o.h.v. engine was evolved, somewhat larger in capacity at 7,428 c.c., and a further stage away from the square dimensions of the early cars with a bore and stroke of $4\frac{1}{2}$ in. and $5\frac{1}{2}$ in. respectively.

The new engine gave more power than its predecessor – 110 b.h.p. at 2,250 r.p.m. – although still unduly modest for its size it was, in fact, stifled by both inlet and exhaust systems. The unit was also appreciably rougher than the Ghost. Lifting the bonnet revealed a power plant of great height festooned with rows of screws, rods, levers and all the usual Rolls-Royce paraphernalia; beautifully made and meticulously fitted, but frightfully untidy in appearance. The notion of combining art with engineering seldom entered the thinking of the design staff at Derby. It is fair to say that the efforts at the Radford works of the Daimler Company were even worse, with the result that the appearance of the V12 Daimlers of the vintage period presented a nightmarish scene, probably only excceeded by the sight of a current V12 air conditioned and fuel injected Jaguar, which by coincidence is assembled in the same plant. An Hispano, Delage or Minerva engine is, by contrast, a model of artistry and tidy, clean design. The Phantom I was subject to various modifications during a four year production run, the most notable probably the adoption of hydraulic shock absorbers, naturally of Rolls-Royce manufacture, and probably the finest dampers in being at the time. The basic design of these units was to remain until the advent of the gas filled telescopic type, introduced with the first Silver Shadow series in 1966.

An aluminium cylinder head was also evolved later in the production run. No doubt this change benefited heat transfer, but was to prove an enormous problem to restorers in later years. Corrosion of the water passages would take place, and a severe case is impossible to repair satisfactorily.

The new Phantom was fitted with front wheel brakes from its inception, together with the well known gearbox drive mechanical servo, itself a refined version of a Hispano design adopted by Derby under licence. The brakes were very powerful by the standards of 1925, but in common with many other makes, axle tramp and shimmy problems were in evidence. Heavier and less precise steering were also penalties, all due to the increase in unsprung weight situated at each end of the axle beam. A valuable accessory also common to the 20 h.p. car concerned manually operated radiator shutters, but mounted vertically in contrast to the smaller model.

The principal British competitors of the Phantom I were the larger models in the Daimler range. The Coventry works had remained faithful to the double sleeve valve engine since 1909, when they secured the English licence to manufacture engines of this type from the American inventor – Charles Y. Knight.

In 1925 the standard large Daimler was a six cylinder 35 h.p. machine of 97 × 130 mm. bore and

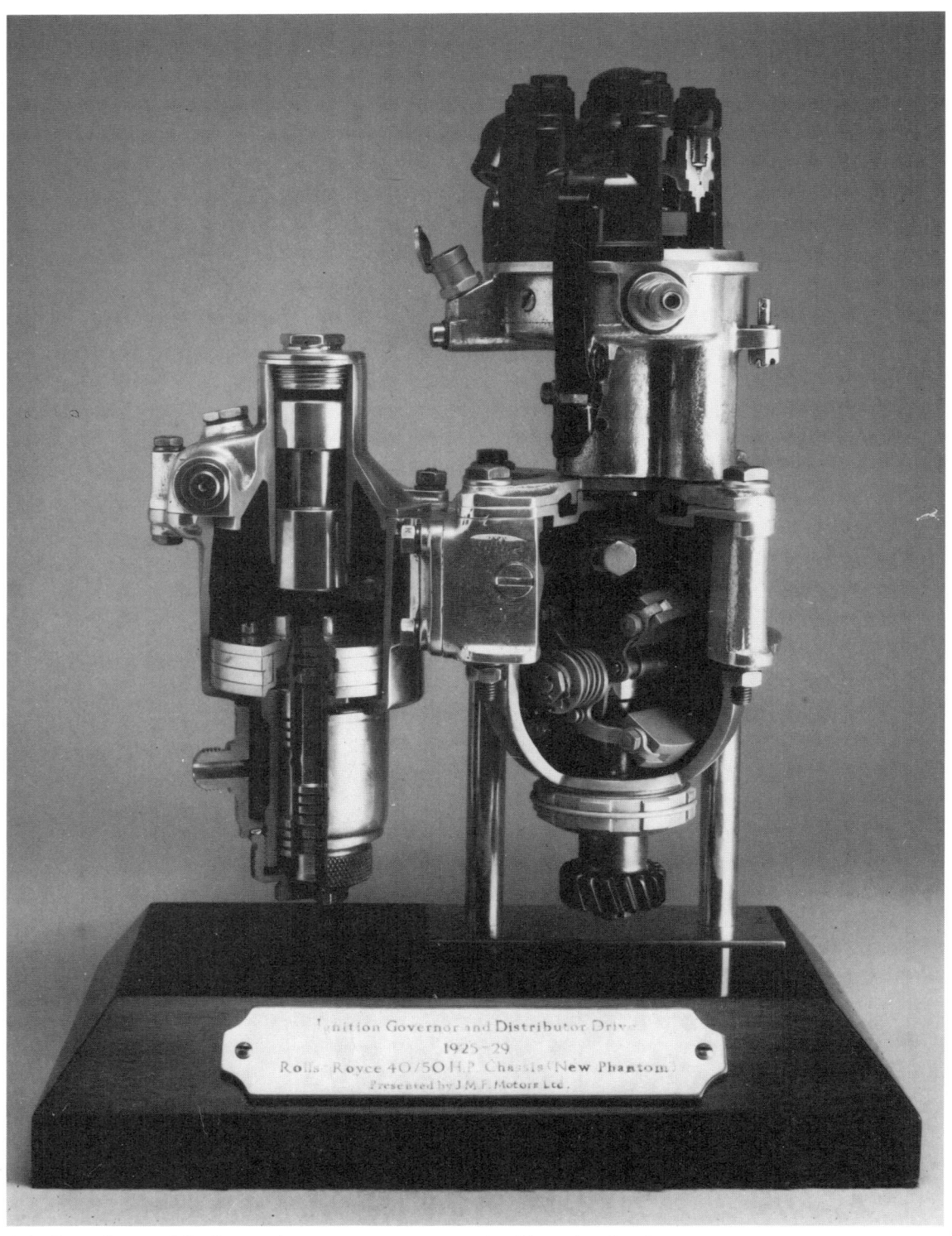

Rolls-Royce Phantom I distributor and governor arrangement, a painstaking and costly refinement.

Phantom I with standard Barker Sedança De Ville body. A 1927 demonstration car retained by sales staff in the South of France.

The original Daimler works, at Sandy Lane, Coventry, in about 1900.

THE DAIMLER CO. LTD. (COVENTRY WORKS)

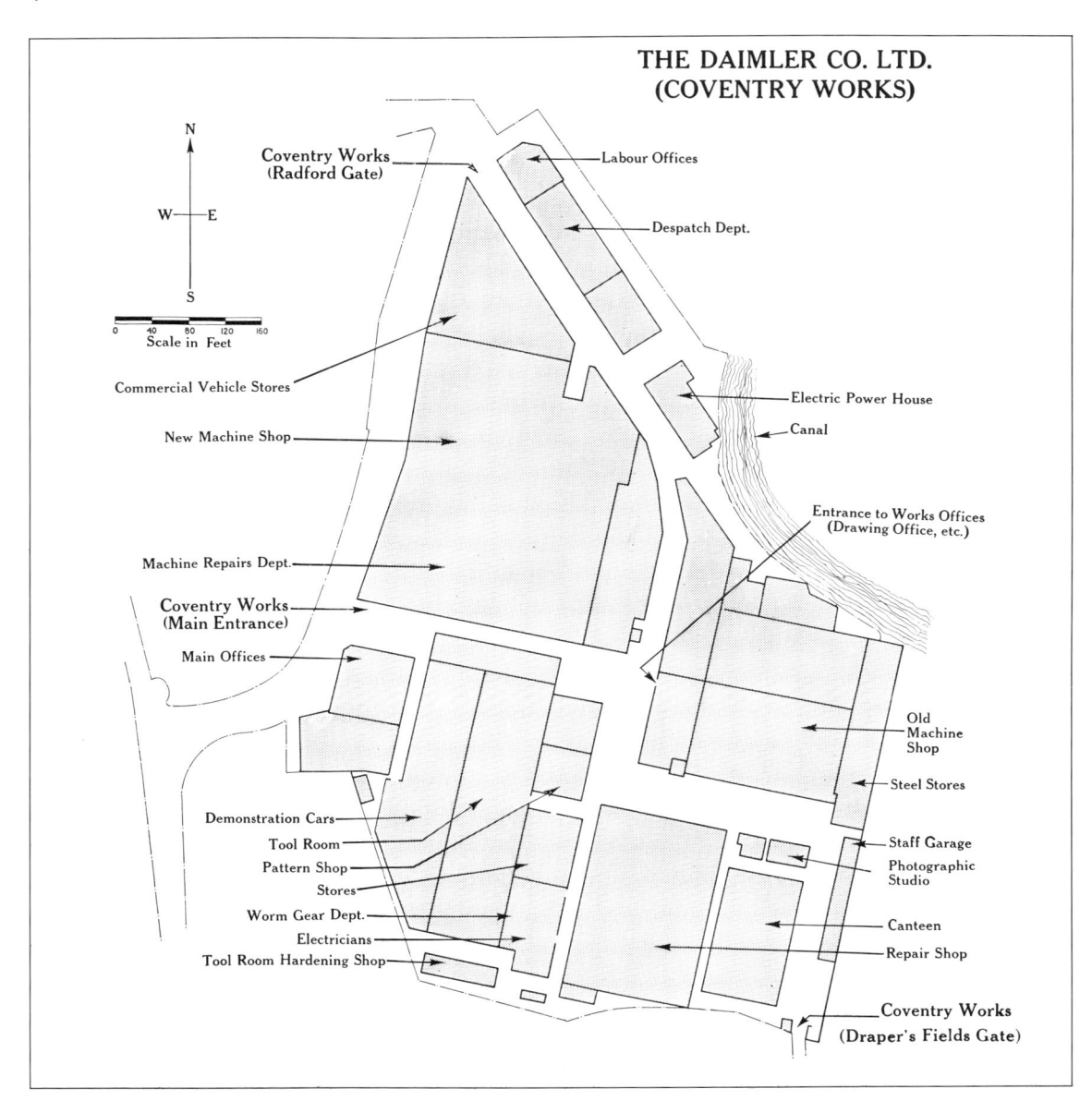

A plan of the Sandy Lane works.

stroke (5,764 c.c.) and thus considerably smaller than the Phantom I and also cheaper, with a £900 chassis price. The performance of this car was distinctly inferior to that of the Rolls-Royce, with a reported maximum speed of 60 m.p.h. An enormous and antiquated six cylinder R.A.C. rated at 57 h.p. was also built in miniscule numbers and, in fact, was mostly reserved for members of the Royal Household, for the Daimler Company were to enjoy continual royal patronage for the first 50 years of this century.

The facilities available at the Daimler works during this period were probably more extensive than those at Rolls-Royce. Two factories within 300 yards of each other were fully engaged on the manufacture of cars and buses. Various other projects were undertaken, including rail car design, while a considerable amount of subcontract work

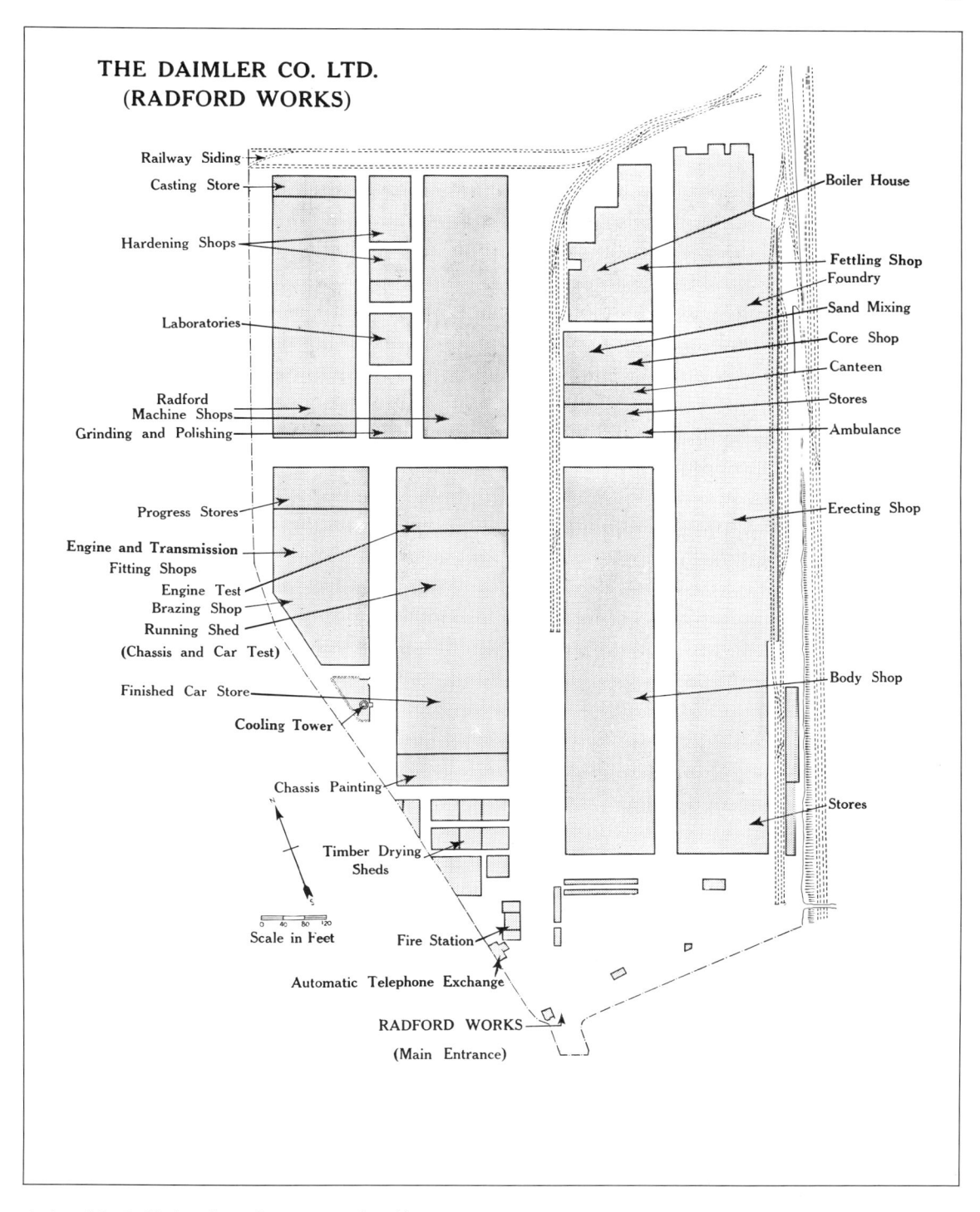

A plan of the Radford works in Coventry, purchased by Daimler in 1908.

A lightweight Phantom I experimental car built in order to assess the possibilities of entering the sports car market. A Schneider Trophy aero engine can just be seen in the background.

was carried out for the B.S.A. parent company. Some twenty draughtsmen were employed on cars although there was a considerable overlap with commercial vehicle work. It is interesting to note that a group of Royce electric cranes were in use in the machine shops and assembly halls.

During 1925 a concerted effort to improve the performance of the private cars took place; a significant development was the adoption of thin steel sleeves, the outer lined with white metal. The result was a considerably lighter assembly than the previous plain cast iron sleeves, resulting in a useful reduction of inertia. Compression ratios were also raised, while ports were increased in size; the outcome was an improved power output, while a new small engine, rated at 16 h.p. and of approximately 2 litre capacity, could be persuaded to revolve at 4,000 r.p.m., apparently without any ill effect. Two other models introduced for the 1926 season were labelled 25/85 and 20/70. The 35 h.p. car, now termed 35/120, gained considerably from the revised design, and examples fitted with light bodies could probably reach 70 m.p.h. Most of these cars, however, were burdened with ponderous formal coachwork. The 35/120 engine was also utilised in bus and coach chassis with great success.

Lawrence Pomeroy, fresh from a tour of duty in America, joined the Daimler concern in 1926 as chief engineer of Associated Daimler Ltd, a firm 50% owned by Daimler and 50% by A.E.C. The objective was the manufacture of lorries embodying the sleeve valve engine, principally the 35/120 type aligned to an A.E.C. chassis. Pomeroy soon interested himself in Daimler private cars and eventually became Managing Director in 1930, succeeding Percy Martin.

Prior to this period, design, development, experimental work, final test and inspection of finished cars was under the control of Lt. A. E. Bush.

The final test, running shed and experimental department were combined in one department. The result was far from satisfactory, and Dr F. W. Lanchester who was still active as a consultant to the Daimler Company strongly advised the splitting of these functions. A company titled Lanchester Laboratories Ltd, 50% owned by Lanchester and 50% by the Daimler Company was then founded, primarily to undertake Daimler experimental work. Due to financial stringencies and a personality clash between Lanchester, Martin and Pomeroy, this project was virtually stillborn.

Pomeroy had achieved wide acclaim for his work at Vauxhall Motors just before and after the 1914–1918 war and was, of course, responsible for the Vauxhall 30/98, one of the finest sporting cars of all time. It was not surprising that he should seek to lift the products of the Radford Works out of the staid and stodgy level in which they had rested ever since the adoption of the sleeve valve engine. Pomeroy was not, however, particularly successful, although Daimler sales continued at a reasonable level until the Depression, which virtually killed their market and forced the development of a cheaper and smaller range of poppet valve cars. This series commenced with an ugly 18 h.p. car bearing the Lanchester name for the 1932 season. The 1926 Daimler range utilised model descriptions which combined the R.A.C. rating with developed brake horse power, a practise common in vintage times, and most useful to prospective purchasers, although subject to abuse and exaggeration from some quarters in later years.

Probably anxiety over lost sales to Rolls-Royce prompted the Daimler management to embark upon a new and dramatic development. The result was an enormous V12 which was announced at the Olympia Show in October, 1926. This car seems to have caused a minor sensation, but despite the interest sales were disappointing, and failed completely to dislodge the Phantom I from its pre-eminent position. The Double Six Daimler, as it was termed, was in effect two sets of 25/85 cylinder blocks, sleeves and pistons placed at an angle of 60° upon a common crankcase. Dual magneto and coil ignition was provided, while twin water pumps were placed under the front of the engine in a most inaccessible position. An external friction damper

Daimler 35/120, 1927. Standard factory coachwork.

Daimler Double Six 50 engine, a nightmare of complexity. Note the drag link flattened out and curved in order to offer an improvement to the right-hand lock. The complicated steering mechanism was necessitated by the width of the twelve cylinder engine.

was fitted to the nose of the crankshaft which looked similar to that fitted inside the timing cover of a Rolls-Royce; both were in fact made under licence from Dr Lanchester.

The induction system was situated, surprisingly, on the outside of the cylinder blocks, and was a tortuous arrangement, consisting of a central riser drawing from an updraught carburettor, which fed into a horizontal pipe through which passed a smaller water heated pipe. The mixture then rose again via two orifices into a further tubular manifold which contained six ports, which led into the cylinders when admitted by the 'windows' in the sleeves. The advantage of a relatively high compression, and efficient combustion chamber shape, made possible by the sleeve valve layout, must have been largely nullified by the stifling induction system. The carburettors were of Daimler design and manufacture, a policy identical to that pursued by Rolls-Royce, but by very few other makers.

The outside induction system adopted for the twelve cylinder Daimler resulted in the exhaust arrangements issuing from the centre of the V. A finned manifold for each bank passed into pipes behind the engine, and down the nearside of the bulkhead. The dual system was retained right through to the rear of the car. A chassis drawing by Gordon Crosby shows extensive lagging of the downpipes, and the problem of insulating the heat generated by over seven litres of engine with such an exhaust system must have been a problem. No doubt most Double Six owners remotely ensconced in the rear of the car were unaware of the plight of the perspiring chauffeur and footman in their separated compartment close to the power unit!

The bulk of this engine precluded the use of a normal steering column, resulting in a high set worm and sector gear mounted on the bulkhead and connected to a normal drag link via a vertical dropshaft and swinging bell crank. Additional wearing points were thus interposed. The remainder of the chassis followed standard Daimler practice, and was offered in three lengths from 12 ft 11½ in. wheelbase, up to 13 ft 7 in. for the full seven seater. A speed of 82 m.p.h. was claimed, while fuel was consumed at a rate of 10 m.p.g. The chassis price was £1,850 for the longer car and thus identical to the Phantom I.

The cubic capacity of the large Daimler was 7,136 c.c. (81.5 × 114 mm.) and 150 b.h.p. was claimed at a speed of 2,800 r.p.m. The Double Six 50 continued in production until 1930, while one or two special orders appear to have been received resulting in the assembly of further chassis at a later date. The only change of note during the model run concerned the adoption of the Wilson pre-selector gearbox, and fluid fly-wheel in 1930. This new transmission was the first commercially successful system offered as an alternative to the conventional sliding pinion gearbox. The eradication of the clutch was also a useful factor which led to its adoption for public service vehicles. It is interesting to recall that the life of the conventional clutch fitted to A.E.C. buses in service with Birmingham Corporation on certain hilly routes was three weeks. A change to Daimler vehicles fitted with fluid fly-wheels resulted in a life of six months under similar conditions before attention became necessary. The part to fail would be an oil seal, much cheaper to replace than a set of clutch plates and linings.

Two special Double Six cars with lowered and shortened chassis were built by Thomson & Taylor of Weybridge, under the direction of a youthful Reid Railton, and while useful publicity resulted from the dramatic appearance of these machines, they were not a success. The length of bonnet in relation to the passenger carrying space probably merits their inclusion in a certain book of records. One was fitted with an open four seater body built to the order of Captain Wilson, who was thus able to test the efficiency of his gearbox for himself, both of these cars being of the later pre-selector series. One survives.

The Radford drawing office, together with the jig and tool department, must have been kept very busy during the late '20s. Apart from a continual flow of new and revised six cylinder cars, a further V12 was laid down immediately after the launch of the Double Six 50. The later car was smaller, known as the Double Six 30, and consisted of two banks of cylinders based on the small 16 h.p. car and thus of 3,744 c.c. (65 × 94 mm). The general design was identical to the 50 although suitably scaled down. Power output was quoted at 105 b.h.p., at the fairly high speed of 3,600 r.p.m. The chassis followed normal Daimler practice, and was available in four lengths ranging from a wheelbase of 10 ft 11 in. up to 12 ft 1 in.; curiously, the two middle length cars varied by a solitary inch, the M type measuring 11 ft 9 in. while the V type stretched to 11 ft 10 in. The smaller lengths were much narrower in track, 4 ft 4 in. against 5 ft, while tyre sizes also varied.

A complete fixed head coupé was listed at £1,570 on the shortest chassis while a full seven seater on the 12 ft 1 in. chassis was priced at £2,180 when announced in August 1927.

The Double Six 30 should have done a great deal of damage to the sales of the Rolls-Royce 20 but, like its larger sister, it failed completely to dislodge the Derby car. It was, in fact, faster and more imposing in appearance as well as enjoying the glamour of a V12 engine. Despite all these sales aids, orders were comparatively slow in coming, which succinctly indicates the innate conservatism of top people in Britain at this period.

Prices were lower by 1929, when a close coupled saloon was listed at £1,300. *Autocar* tested such a car in February of that year, which they coaxed up to 75 m.p.h. while waxing enthusiastic about the unique silence and flow of ample torque right through the speed range.

The Double Six 30 remained as a listed model

Daimler Double Six 30. A 1929 car in an Oxfordshire setting, coachbuilder unknown. Note the Newton telescopic dampers and early form of bumper bars, a proprietary fitting.

through the 1931 season, although it is doubtful if many were assembled after 1930, for yet another V12 had been evolved by that date. In 1929 Lawrence Pomeroy instigated the design and production of an advanced six cylinder car of similar capacity to the previous 25/85 model. This new machine, simply titled 25, was announced in October and thus was ready for exhibition at the Olympia Show. The car was aimed at the owner driver and a determined attempt was made to project this new Daimler as a high performance sports saloon. Pomeroy had, of course, been associated with high performance vehicles in his work at Vauxhall Motors, and the Daimler 25 project no doubt appealed to him. He had a great enthusiasm, perhaps amounting to a mania, for the use of light alloys in motor car design. He did, in fact, carry out a design project in the U.S.A. circa 1922 for a car made almost entirely from this material, for the Aluminium Corporation of America. The resultant vehicle resembled a Vauxhall even to the extent of incorporating flutes in the bonnet.

The Daimler double sleeve valve engine underwent a further and final re-design for the new car, and incorporated a single monobloc casting for all cylinders of aluminium, while a daring feature concerned the fact that the sleeves oscillated directly in the aluminium bores. The separate cylinder heads and one piece enclosing water jacket for the above were also of aluminium, while the pistons were of the recently introduced invar-strut design which were to enjoy a vogue for some years. The object was a reduction in the rate of expansion of a normal aluminium piston, thus allowing tighter clearances, a laudable aim for a sleeve valve engine with its voracious appetite for oil. The engine was of a much neater appearance than the previous model. The induction system would appear to be the least attractive feature and consisted of a single carburettor, once again of Daimler manufacture, although improved in design. This instrument was situated on the nearside front of the engine, close to the exhaust manifold, which was jacketed to provide a warm air intake, the degree of heat intake being adjustable. The inlet manifold then passed around the front of the engine into a single long pipe along the offside of the engine incorporating separate ports for each cylinder. It would seem unlikely that the mixture strength for all the cylinders would be equal. 87 b.h.p. was claimed at 3,600 r.p.m. The chassis followed traditional Daimler practice although light alloy components were introduced wherever possible, even for the front axle beam. This car must have been costly to produce. The use of light alloy for load bearing components such as a front axle requires the use of a sophisticated alloy, forged and carefully heat treated if a good safety factor is to be attained. It is significant that this extravagant method was never repeated on subsequent models. We now learn that, in fact, failures of the front axle did occur.

Silentbloc spring shackles were also adopted for the 25 together with the new Luvax vane dampers in place of the Newton hydraulic pattern used previously. The vane dampers proved to have a woefully short life before fluid leaks became apparent. Nevertheless, they became an almost universal fitting on British cars for several years. It is to the credit of Rolls-Royce that they never fell for this one. The 25 was listed complete with Weymann sports saloon body at £1,250, or £1,050 with slightly more angular coachwork by Arthur Mulliner of Northampton. The chassis was priced at £700. The writer's family ran one of these Daimler 25 cars for a short period before the war. It is remembered as an imposing machine fitted with a close coupled Weymann Saloon body in blue and grey fabric, and fitted with cycle-type wings and oval step boards below the doors, of a type once fashionable. It was considered one of the smartest cars in Warwick at the time and embodied the pre-selector transmission, standard for the second year of production. It was quite fast, but the feature of silence which was the raison d'être of the Knight engine was only evident at tickover. A considerable clatter could be heard at higher speeds, which emanated from the sleeve links which had become worn after a considerable mileage. The car also consumed both oil and petrol at a frightful rate, and was eventually scrapped. Such a car, despite the fine

Daimler 25, 1930. This is the lightweight model in standard form, with factory-built fabric coachwork on the Weymann principle.

Daimler 30/40, 1931. An example of the second generation of V. 12 cars from the Bradford works. Thermostatic shutters feature in a re-profiled and deeper radiator.

condition of the body, was totally unsaleable in 1938.

The 20/25 Rolls-Royce was introduced at the same time as the Daimler 25, and despite the lower price, better looks and performance of the latter, orders taken were, once again, at a much lower rate than those enjoyed by Derby. The car was discontinued in 1932, by which time prices were £700 for the chassis, while a complete sports saloon was listed at £975, indicating a clearance price for the complete car. A smaller version of the revised six cylinder sleeve valve engine rated at 20 h.p. (2,648 c.c.) was introduced in the early part of 1930 and this car, which continued in production until late 1933, was to form the basis of yet another V. 12.

This machine, titled Double Six 30/40, was announced in September 1930 and comprised two banks of the 20 at the usual angle of 60°. The dimensions were 73.5 mm. bore × 104 mm. stroke (5,296 c.c.).

The engine was considerably simplified compared with previous twelve cylinder models and was much tidier in appearance. A single water pump sufficed while the ignition system now relied on a dual coil system, and the fuel feed was by A.C. mechanical pump. The exhaust manifolds were still positioned in the centre of the V, although the offtakes were now positioned at the front of the engine. The induction arrangements remained similar to the 20 and 25 models with forward mounted carburettors, one for each bank and still of Daimler manufacture. The 30/40 received the fluid flywheel and pre-selector gearbox transmission from the outset. A comparatively high final drive ratio of 3.78 to 1 was specified, which gave a road speed of 24½ m.p.h. per 1,000 r.p.m. The total weight of a touring limousine version on the short chassis at 51 cwt. held down acceleration. Maximum speed was 75 m.p.h. A new and extremely low chassis was evolved which resulted in good handling and a better looking car, even in limousine form, than the majority of formal Rolls-Royce bodies of the period.

The Daimler braking system left something to be desired. The distinctive Daimler wheel with centre locking cap embodied a very large hub which completely shrouded the brake drum. This feature also precluded effective finning on the drum, as well as putting a severe limitation on diameter. The operation of the braking system remained conventional, with a well-thought-out system of rods operated by levers and cross-shafts now running on roller bearings. The Clayton-Dewandre servo was retained. The short chassis was listed at £1,100 with complete saloons costing £1,475 and upwards.

The Double Six 30/40 fell exactly between the two Rolls-Royce models; once again sales figures were very poor. It is amazing that Pomeroy and his directors should resolve upon yet another V12, adding an enlarged version to the range for 1931. This version consisted of two banks of 25 h.p. cylinder blocks and probably utilised many parts common to the 30/40, and so the development cost of this final version may have been minimal. Dedication to the V12 engine configuration at the Radford works did not finally expire until a poppet valve version had been tried in 1936–7. At least two cars were so built, and one is still in the Royal Collection.

The fact remains that, depite strenuous efforts, the large Daimlers of the late vintage period failed to dislodge the firmly entrenched position of Rolls-Royce. The Coventry cars were built to exacting standards with finish and detail work, perhaps the equal of that obtaining at Derby. The palm for innovation and advanced thinking must also go to the technical staff at Daimler.

One comes to the conclusion that the inherent difficulties with the Knight engine, principally concerned with oil consumption – which was heavy when new and absolutely chronic after wear had taken place – was a major drawback. It follows that the resultant blue haze was offensive.

Towards the end of the sleeve valve era, Lawrence Pomeroy Jnr came back from the U.S.A. with a concession for a new form of flexible scraper ring which, after further development, cut oil consumption down to virtually nil. Some blue smoke remained, however, which was thought by some to be oil passing through the exhaust port. This neat oil would presumably burn while passing through the exhaust system, hence the smoke.

The Daimler board also decided, in view of the depressed market in 1931, that they should embark on the production of greater numbers of cheaper cars. The sleeve valve engine, with its need for highly skilled fitters when service became due, was thought, quite rightly, to be a drawback for the enlarged market envisaged.

With the benefit of modern materials and lubricants there may, however, be a case for taking a further look at the possibilities of a sleeve valve engine layout.

CHAPTER 5

1930–1935
The Phantom II and the 8 Litre Bentley

When one considers the last big six cylinder Rolls-Royce current from 1929 to 1935 and steadily developed throughout this period, it is somewhat difficult to compare with other luxury cars. Perhaps the obvious contemporary car would be the V12 Hispano-Suiza, announced some twelve months after the Phantom II, current for roughly the same span and which came in two sizes. It would also be interesting to make a comparison with a high grade American car. There were several excellent contenders including the K series V12 Lincoln, the Senior Packards, the rather similar Pierce-Arrow and the General Motors flagships in the form of V12 and V16 Cadillacs. Most of these imposing machines were manufactured in mass production quantities compared with European equivalents, quantities which reveal the enormous prosperity of the United States, the Wall Street crash not withstanding. The 8 litre Bentley would seem to be the obvious British car which was intended to offer a serious challenge, and the absorption of Bentley Motors by Rolls-Royce in 1931 resulted in an instant shut down of the production of this fine car.

The Phantom II retained the engine dimensions of the Phantom I but was subjected to a substantial re-design. The inlet manifold now appeared on the offside of the unit, while a neater three-branch exhaust manifold carried waste gases from the nearside. The bottom end of the power plant remained largely unchanged, but the gearbox now appeared in unit with the engine in the manner of the smaller cars; indeed, the whole chassis resembled a scaled up 20/25. Throughout the model run changes were made to the specification in accordance with Rolls-Royce practice. The most important concerned the addition of synchromesh to the gearbox and ride control to the shock absorbers. Various changes were made to the cam profile from time to time, no doubt brought about by anxiety over noise; while satisfactory at tickover, these large engines were apt to make a loud commotion at high speeds. The fan was also a major culprit.

The 8 litre Bentley followed the standard practice of the Cricklewood firm in that the six cylinder overhead camshaft engine was a legacy from 1914–1918 aero engine practice. The four-valve-per-cylinder layout allowed very good breathing while the excellent combustion chamber shape conbined in unit with the cylinder block allowed good heat transfer and cooling of the exhaust valves. This design obviated one of the main problems of the Phantom II engine – which admittedly did not become apparent until later in life. This was the serious electrolytic corrosion, mentioned earlier, which affected the aluminium cylinder head. The power output of the Bentley was probably 50 per cent greater than that of the equivalent 1930–1931 Phantom II and the road performance was therefore vastly superior.

Tickover silence was almost as good as the Phantom II, although the enormous engine made the usual roar at higher speeds, while the exhaust system was allowed to make its presence heard – probably acceptable in a car bearing the Bentley

Rolls-Royce Phantom II, 1933. Nearside of 105 MW, relatively neat and tidy by Derby standards.

name and affording a reduction in power loss from this source in comparison with the Derby product. The Bentley engine, while capable of higher revolutions, was also smoother at the top end, probably due in part to the stiffer engine construction. It is amazing that Rolls-Royce continued into 1935 with the Phantom II. The power unit was, by then, distinctly dated and outrageously rough when compared with any of the better class of American or European cars. To drive one of these large cars at speeds in excess of 70 m.p.h. is definitely not a pleasure. It is generally agreed that pre-war road tests by the motoring journals are sometimes open to question, and it is a source of utter amazement to the author that *Autocar* managed to achieve a mean speed of 94 m.p.h. with a saloon example in 1934.

The roadholding and steering, the latter allowing severe kick back through the steering wheel, is very disappointing, while alarming axle tramp will manifest itself unless the front shock absorbers are fully effective. The steering is, however, extremely light, providing the king pin lubrication is in good repair. Rolls-Royce practice in pre-war days was to utilize hardened steel stub axle bushes instead of the customary bronze; this material was quite satisfactory in service provided the one-shot oiling system was kept in good order. The Bentley steering, although superior at speed, was definitely heavier in operation. The ride, especially for rear seat passengers, was poor in the case of both makes, the Bentley in particular. Again, any American car was much better in this respect.

The Rolls-Royce gearbox was easier to use than the Bentley, indeed the later synchromesh examples were delightfully silky in operation, although gear changing was relatively slow. The Bentley box was quite tricky, but once mastered, would prove a delight to its driver. Both designs were immensely strong pieces of mechanism.

The two cars were equipped with very powerful braking systems, although the vacuum servo used by Cricklewood suffered from the inherent disadvantage mentioned earlier.

Offside front of 105 MW. The cylindrical unit in the foreground houses the oil pressure-operated advance and retard mechanisms.

A maharajah's stable. The dramatic social upheaval that has taken place in India is vividly portrayed by the crumbling buildings and delapidated motor cars seen here in about 1975. From left to right: Lanchester 40, Rolls-Royce 20, Phantom II 105 MW and two Silver Ghosts.

Phantom II 105 MW, after restoration, in Stratford-on-Avon Museum.

Phantom II, 1933. Barker sports saloon, chassis No. 104 MY. Originally reserved for the use of A. F. Sidgreaves, Managing Director of Rolls-Royce Limited, later owned owned by the author, the performance was a little faster than standard. The interior incorporated silver-plated window fillets and simulated lizard skin upholstery to front.

A Phantom II, 1933. Thrupp & Maberly sports saloon.

Bentley Motors followed the Derby practice in that coachbuilding was not included among their facilities and both firms relied almost exclusively on the houses of the leading London coachbuilders. The aesthetic appeal of motor cars reached its zenith in the early '30s and some of the later Phantom II chassis, fitted with sports saloon or coupé coachwork, are amongst the most elegant cars ever built, body design being aided by the enormous bonnet length and high waist line.

The overall reliability of the Phantom II, as with all Rolls-Royce products, was extremely good, largely due to the magnificent quality and design of all the electrical equipment, normally the most troublesome area of a motor car. Alas, the quality of this equipment was soon to be sacrificed in the interests of cost reduction.

The Phantom II was conventional in design, although one or two special features not seen elsewhere were in evidence. One such item concerned the exhaust heated inlet. This utilised a cast steel jacket interposed between the carburettor and manifold. This system worked well but the casting suffered from internal corrosion. The free space inside the jacket was very limited and the expansion of the walls, when heavy scaling built up, completely blocked the passages. This process would eventually burst the casting. This scheme was given its own separate exhaust comprising a small pipe and silencer system – not a very good arrangement, adding weight, complication, cost and another corrosion-prone set of parts.

An 8 litre Bentley with standard specification power unit

CHAPTER 6

1929–1936
The Rolls-Royce 20/25 and the Hispano Suiza H.S26 and 30/120

The second model of the junior cars from Derby was current from 1929 until 1936 and was, in the opinion of the writer, the most attractive of them all. The old 20 h.p. engine received a bore increase of $\frac{1}{4}$ in. (3,669 c.c.) together with various minor improvements. Approximately 80 b.h.p. was developed at 3,400 r.p.m. The new car possessed a much better performance than the previous 20. It was quite lively, except when burdened with heavy limousine coachwork, while the handling and general feel was good; but, once again, the car became less pleasant at higher speeds, with the usual vicious Rolls-Royce kickback through the steering wheel. Development through the years continued, the adoption of synchromesh being probably the most notable. Unfortunately weights tended to increase, and the charm of earlier models was lost to a degree by the time the model was superseded by the $4\frac{1}{4}$ litre 25/30 model in 1936.

The direct British competition during the first two years of production came from such cars as the Lanchester Straight 8 mentioned previously, together with the medium size Daimler range, in particular the 25 h.p. The larger Sunbeams struggled on until mid-1935 before bankruptcy finally intervened. Their 20 was enlarged to 3,317 c.c. (R.A.C. rating 23.8 h.p.) and was given seven main bearings in 1931 while Lockheed hydraulic brakes were adopted for 1932, and eventually synchromesh gearboxes; these cars still remained superior in performance, and were beautifully built right to the end. The Wolverhampton firm also revived the old 20.9 h.p. unit, gave it an improved downdraught induction system, and then evolved an extremely smart sports saloon for the 1933 season. A fast car resulted which was intended to compete with the recently introduced Speed 20 Alvis, indeed Sunbeam adopted the same title. Perhaps this car, which underwent a further engine change for the final 1935 season to be known as the Speed 21, would have made inroads into the $3\frac{1}{2}$ litre Bentley market if the business had survived. It is interesting to note that the Sunbeam body engineers had adopted cast aluminium windscreen pillars for this car in a similar manner to that of Park Ward on the Bentley.

The European opposition to the small Rolls-Royce was spearheaded by the small Hispano Suiza which was now designed and built at the Ballot factory in Paris. This new model, titled H.S.26, superseded the previous 27 h.p. car which was made in the Barcelona works, and resembled a scaled down version of their well tried 37.2 h.p. H.6.B car. This latter vehicle had always been built in Paris and had endured since 1919 with only minor revisions. The Ballot-Hispano appeared in 1931, and was said to consist of a Hispano engine which was fitted in the chassis of the Straight 8 Ballot, itself a short-lived model.

An inspection reveals, however, that the H.S.26 chassis followed the layout of the H.6.B very closely, and many items would appear to be interchangeable. The chassis frame was entirely conventional, and almost totally lacking in cross members; the dumb irons front and rear were braced by cross tubes of approximately $1\frac{1}{4}$ in.

A 20/25 from 1929, with Weymann coachwork.

diameter. The first cross member appeared approximately 3 ft 6 in. ahead of the rear axle line, and carried the torque tube nose. This latter component was, therefore, comparatively short, and necessitated an open propeller shaft with a flexible coupling at its forward end to transmit the drive from the gearbox. This arrangement was similar to that adopted by the Austin concern for the 7, but was unusual at the time. It had been seen in modern times on Riley cars (1946–1955) and the current S.D.I. Rover range. The Hispano chassis was braced by a further cross member just aft of the rear axle, while the entire front end stiffness was provided by the engine crankcase, a very rigid casting secured to the chassis at four points. The long semi-elliptic springs with well tapered and ground leaves (nine front and twelve rear) utilised Silentbloc rubber bonded bushes throughout, unusual on a European quality car at that date, although well established in America. The main leaves on the Hispano curved around the eye, then forward along the spring for 2 in., the ends secured by two rivets, again American practice. The rear axle consisted of a deep flanged steel banjo pressing; the axle design appears entirely

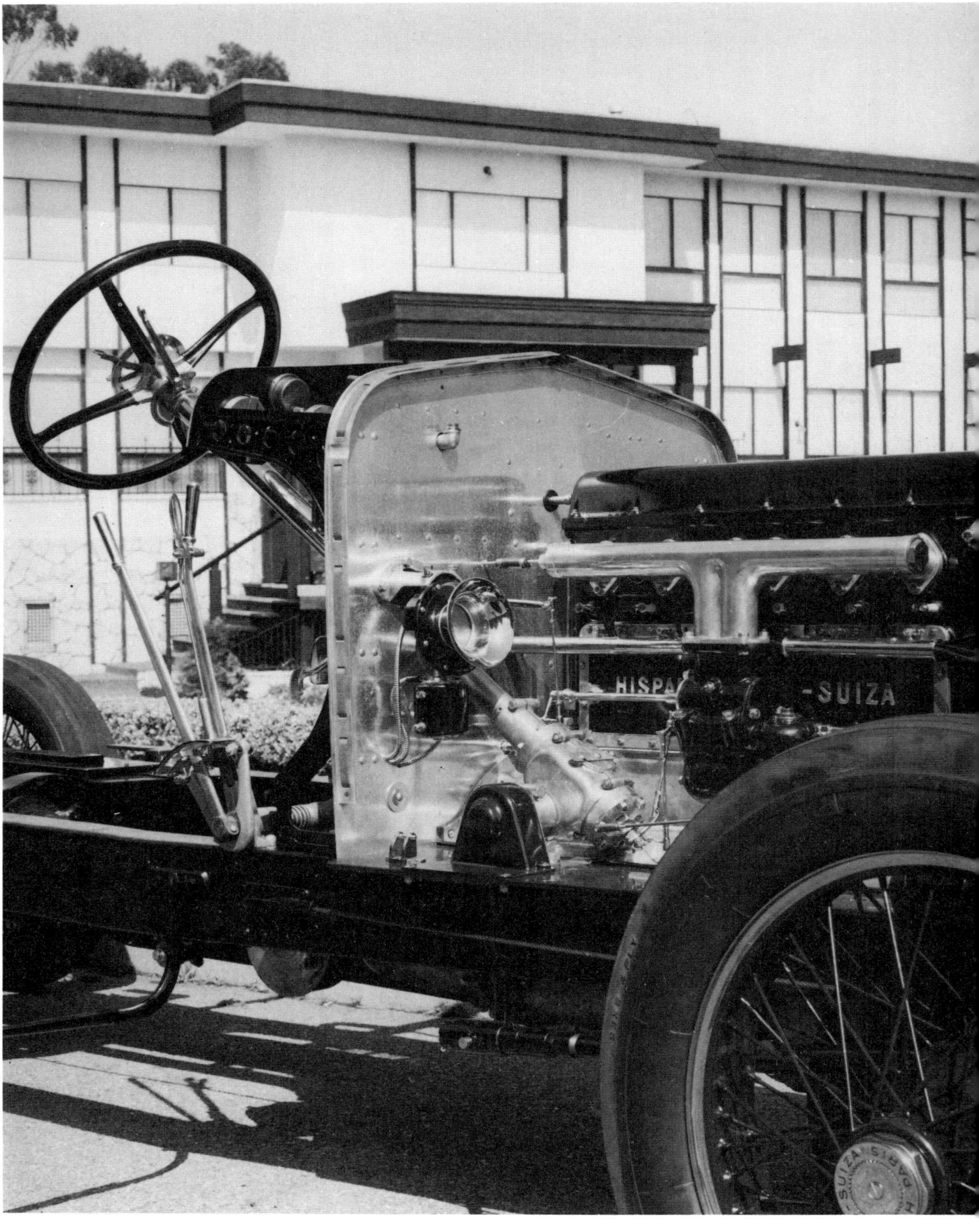
HISPA
-SUIZA

Hispano Suiza H.6.B, showing the engine finished in standard Hispano black stove enamel and polished aluminium.

conventional, and the torque tube is secured to the differential and pinion housing by 14 studs of approximately 11 mm. diameter. The axle was secured to the springs through a trunnion mounting, necessitated by the torque tube layout.

The front axle was a conventional I section steel-forging, but appears to be machined on almost all faces, and the spring pads are gracefully tapered into the beam. Hydraulic shock absorbers of Houdaille manufacture were specified, not unlike Rolls-Royce units in appearance.

Two items of Hispano invention are to be found, both taken up by Rolls-Royce Ltd, namely the serrated hub nut locking mechanism (via Dunlop), and the mechanically operated brake servo motor. The hub nut used on the H.S.26 is smaller in diameter than that of the 20/25 Rolls-Royce; the French firm used exactly the same size on their $6\frac{1}{2}$ litre H.6.B model, upon which they appear distinctly mean. Those on the V12, however, are enormous.

The braking scheme adopted was of the Perrot system, then so popular, and operated by cables in the case of the Hispano with a neat little differential gear interposed as a compensator. Enormous brake drums of approximately 16 in. diameter were provided and the brakes were considered exceedingly powerful, with the good feel associated with mechanical servo systems.

All Hispano Suiza cars were fitted with three-speed gearboxes. The H.S.26 unit comprised a cast iron case, small at 7 in. in total length, while the detachable bell housing and tail casting which carried the right-angle servo drive were of aluminium. The constant mesh input pinions, together with first and reverse, were straight cut, although the second speed gears were of single helical form. A centre ball change was provided.

This unit also reflected American practice and appears somewhat coarse compared with the beautiful mechanisms that hailed from Rolls-Royce. It is fair to say that the entire Hispano chassis appears disappointing in comparison with the other luxury cars described in this book. The power plant is, however, a totally different matter. Yet another

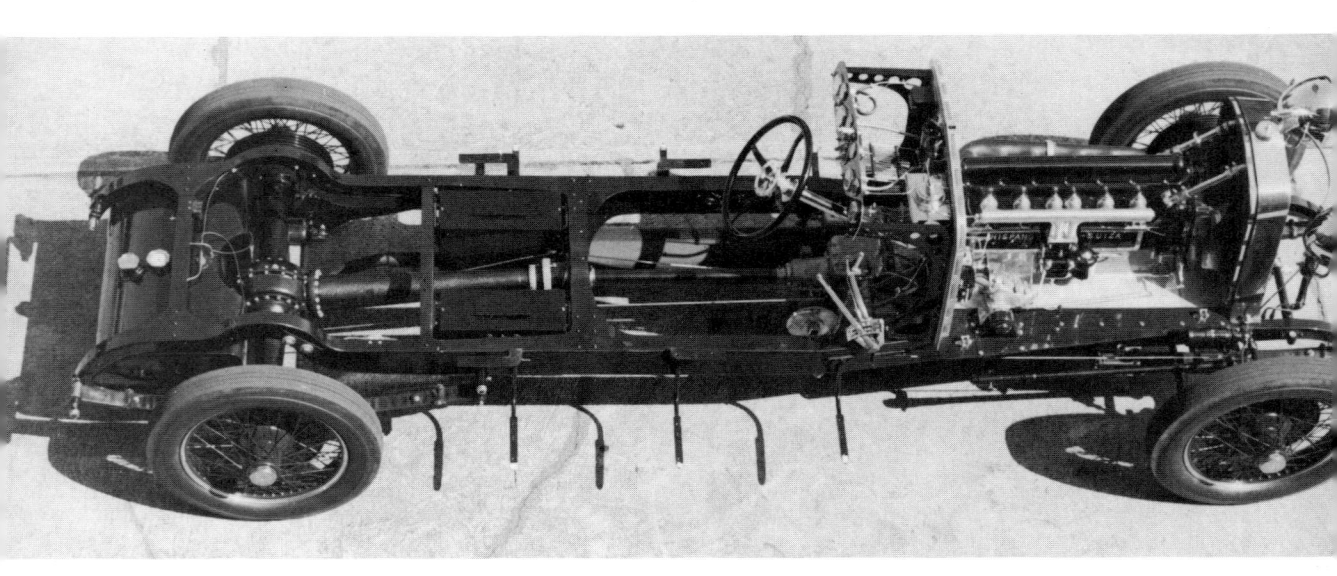

Hispano Suiza H.6.B chassis. Note the well-proportioned gussets joining chassis side rails to cross members. The tyres appear under-sized on this restored chassis.

descendant from 1914–1918 aero engine practice, it was, in fact, a scaled down edition of the H.6.B; the reduction in size from 6½ litre to 4½ litre only resulted in a loss of 2 in. in overall length, and 1½ in. in height. The whole power plant is extremely neat and clean in design, and is a production of the highest grade.

Six cylinders of 90 mm. bore with a 120 mm. stroke result in a capacity of 4,560 c.c.; it is, therefore, considerably larger than the 20/25 Rolls-Royce. The crankshaft ran in seven bearings and the crankcase casting embodied very wide full length flanges which fitted snugly up against the chassis frame. It made for an extremely rigid engine, as well as imparting much needed torsional stiffness to the chassis.

A dry five-plate clutch of Hispano design was incorporated, while a very large Scintilla dynamo was driven off the nose of the crankshaft and protruded through the radiator; a polished aluminium cover incorporating a screwed cap to conceal the starting handle dog was provided.

The aluminium cylinder block and head was cast in one, with wet chromium liners screwed into the combustion chamber. Ample water passages are apparent. The overhead camshaft was driven from the front by a vertical shaft and bevel gearing. The cooling system was assisted by a water pump located on the nearside front of the crankcase, and manufactured entirely from brass, finished by chromium plating. Two water outlet pipes took off from the front of the cylinder head.

The inlet system comprised an updraught carburettor of joint Hispano-Solex design which fed into five inlet ports, the central cylinders being siamesed. The carburettor drew air from a large box cast into the offside crankcase flange previously mentioned, and two air intake apertures were provided in the bottom of the box. A neat distance piece of cast aluminium took up the space between the carburettor air intake and the crankcase, secured by a spring clip. A six-port exhaust manifold was provided with a forward offtake, the manifold passing through the nearside crankcase flange. The exhaust pipe to manifold four bolt flange is, therefore, to be found underneath the car. The clean underbonnet appearance of the larger H.6.B model, which utilised a shorter manifold casting with the pipe flange in view, was thus enhanced still further with the H.S.26.

The cylinder block and valve cover were well finished in black vitreous enamel, a Hispano feature. A pair of Scintilla Vertex magnetos were situated on each side of the camshaft drive shaft, each feeding a row of plugs on their respective sides of the engine. A very neat advance and retard mechanism was provided. This system was concealed inside the front gear cover, and presumably embodied some form of scroll incorporated in the magneto drive shafts.

The power output was in the region of 95 b.h.p., which provided a performance considerably superior to that of the 20/25 Rolls-Royce. The only car of this type resident in the British Isles has been in the ownership of the Earl of Moray for close on thirty years. It is fitted with an attractive drophead coupé body, the work of the North London firm of Lancefield. Lord Moray feels that the performance suffers from the fact that it appears undergeared, and is thus felt to be working hard at speeds above 55 m.p.h., while the three-speed gearbox is probably the most serious drawback. The fuel consumption averages 15 m.p.g., whereas it is possible to achieve 18–20 m.p.g. with a 20/25 Rolls-Royce. The valve gear on this Hispano is also noisy, which probably accentuates the general feeling of commotion at higher speeds. The owner is, however, most impressed with the roadholding, steering and general feel, which is that of a thoroughbred.

The Hispano certainly failed to sell in serious quantities, and production probably totalled about 100. Lord Moray's car, known to be a late example, is considered the sixty-sixth built. The year 1935 saw the introduction of a new, small Hispano known as the 30/120. This car, virtually identical in chassis layout to the H.S.26, was provided with a new power plant of 100 mm. and 110 mm. bore and stroke (4,900 c.c.). This unit consisted of one bank of the V12 engine which itself had replaced the H.6.B some years before. The stroke, at 110 mm., fell half-way between that of the two sizes of V.12 that were offered. Initially, these enormous engines were of square dimensions (100 × 100 mm.), 9,424 c.c.

In order to improve further the massive torque available a lengthened stroke version of 120 mm. (11,310 c.c.) was offered. This later range of engines

Hispano Suiza H.6.B instrument board, with its large Nivex petrol gauge on the left.

The zenith of elegant design during the art deco period is represented by this Hispano Suiza V.12 drophead coupé, circa 1935.

embodied pushrod overhead valve gear, and the clean design and beautiful finish was once more very much in evidence. The new six cylinder car was offered in two wheelbase lengths, and the gearing was now reasonably high, at 3.65 to 1 for the long chassis and 3.4 to 1 for the shorter model. A long chassis saloon was imported as a demonstrator by the British concessionaires and proved capable of 82 m.p.h., together with highly respectable acceleration. The short chassis cars, when fitted with light coachwork, were said to be good for 90 m.p.h., a performance which was enormously superior to that of the smaller Rolls-Royce. The Hispano was listed at the substantial price of £1,895 when fitted with a pillarless saloon body by the Parisian coachworks of Van Vooren.

There is no doubt that the Hispano Company were distinctly half-hearted in promoting their car in export markets. The British concessionaires, Messrs. Jack Smith of Albemarle Street, W.1., were given a simple folder, badly printed with the barest of facts, to hand to prospective clients. The beautifully produced large catalogues issued by Rolls-Royce, Daimler and, to be fair, by the expansive Americans, were in stark contrast. Minerva, almost alone among the European manufacturers, spent lavishly on the production of tasteful catalogues.

The Hispano Suiza of the 1930s was undoubtedly one of the world's most beautiful and glamorous cars, invariably fitted with extravagent French

Engine of the rare Spanish-built four cylinder Hispano Suiza Type 48, from about 1929. It is almost identical in layout and finish to the large Paris-built cars.

coachwork in the Art Deco manner. Nevertheless, sales figures never remotely approached that of Rolls-Royce; the firm lacked aggressive marketing, and the Directors concentrated the emphasis, no doubt wisely, upon the well-established aero engine business. It is interesting to note that the top car firms of England and France developed a large and highly successful aviation power plant business in the First World War, which continues until the present day; when both firms faced bankruptcy in modern times, their economic importance was such that rescue operations by the state ensured their survival.

CHAPTER 7

1934-1939
The Bentley $3\frac{1}{2}$ and $4\frac{1}{4}$ Litre and the Bugatti Type 57

Rolls-Royce purchased the assets of Bentley Motors Ltd in November, 1931. Some initial vacillation took place over the type of car which Rolls-Royce would build to bear the Bentley name. The keen young men at Derby were pressing for a super sports car on the lines of an 8c Alfa-Romeo. Sidgreaves, the Managing Director, and Royce were, quite sensibly, not disposed to take this course. Sidgreaves pressed the Derby engineers to take a look at the Lagonda and Invicta which looked 'boldly British and our type of car'. He also suggested that design work would be saved if they took half of a Kestrel aero engine and scaled it down. 'You know it is capable of taking a supercharger', said Sidgreaves, indicating another line of thought which was occupying the minds of most sports car manufacturers in the early '30s.

The final specification sanctioned for production was undoubtedly the best course and, once decided upon, work proceeded with commendable despatch; the first experimental car was handed to the Derby test department in January 1933. Production cars were ready for demonstration to the press six months later, although proper deliveries did not commence until the beginning of 1934.

It is known that design work for a small Rolls-Royce car of 16–18 h.p. (R.A.C. rating) had recently been undertaken, known as the Peregrine project. Advantage of this work was taken while a

The first experimental $3\frac{1}{2}$ litre Bentley completed in January 1933. Note the eared hub caps, swage lines on the wings and louvred windows, all abandoned on the production cars. The bonnet louvres are vertical.

An air line saloon to the order of Geoffrey Smith, managing editor of Autocar.

An experimental 18 h.p. Rolls-Royce, 1932. Note the Rudge wire wheels.

new lowered frame was evolved and Rudge hubs were adopted; the engine was based on the 20/25 unit. A new cylinder head and induction system were evolved, resulting in a power improvement of some 50 per cent. The lower and lighter car was obviously going to be capable of 88–90 m.p.h. and would suit the performance-conscious Bentley customer; he would be able to enjoy Rolls-Royce delicacy and refinement as a bonus. Quite a number of these diehards were ungrateful enough to trumpet loudly that the new car was effete and far too soft. Nevertheless, the 3½ litre Bentley sold well from the start and was undeniably an attractive proposition.

Rolls-Royce were now dominating the expensive car market, while the size and strength of the company far outstripped that of their most serious British contender in the large sports car field,

The two original 3½ litre cars photographed on their first public appearance in July 1933, probably at Brooklands. The Saloon is still extant, owned by an enthusiast in Leicestershire.

The first production saloon. Note the sloping bonnet louvres revealing very well-balanced and clean lines, enhanced by small headlamps and a lack of bumpers.

A photograph of ALU 323 from the previous picture, dilapidated after 50 years and awaiting restoration

Artist's impression of the original 1934 Van Den Plas tourer.

namely Lagonda. The European opposition had also contracted. As mentioned in the last chapter, Hispano Suiza had decided to abandon motor car manufacture by the late '30s and in any case did not contest the sports car market.

The last serious model from the fabled Alsace works at Molsheim, the Type 57 Bugatti, spanned exactly the same period as the Derby Bentley, was similar in size and aimed at exactly the same market, namely persons of discernment who had sporting instincts and who sought the ultimate in quality. This factory from the east of France, although well equipped, was by now mainly engaged in the manufacture of petrol engined rail cars for the French state railways. They were unbelievably weak financially, and from the middle '30s were under the direct management of Jean Bugatti who, in 1934, was a mere 25 years of age.

Design work on the Type 57 was carried out in 1932, and the first car appeared with independent

The 1936 Van Den Plas tourer. The wing and running board line up is ill-judged. Chrome mouldings have appeared on the waistline neatly blended in to the windscreen stanchions.

front suspension incorporating transverse leaf springs, one above the other, a system adopted by the Bugatti works for two experimental four-wheel drive hill climb cars in the previous year. It is just possible that the Alsaciens may have noticed the similar arrangement used by Captain Smith-Clarke for his front wheel drive Alvis cars. B.S.A. also adopted this layout for their f.w.d. light cars – vehicles never noted for stability. The scheme has an inherent drawback due to a high roll centre.

The only Type 57 built with this feature appeared at the Swiss Motor Show in 1933. It was hastily withdrawn at the behest of Ettore Bugatti, who insisted that the new car should follow standard Bugatti practice, which consisted of a solid front axle of most artistic form, almost straight, and incorporating boxes through which the front springs passed, thus retaining rigidity and retention of the designed steering geometry under braking as far as possible with a non-independent layout.

The production cars, of which the first example is owned by Ross Harper of Solihull, was a larger car than the previous Type 49 model, although the engine dimensions remained similar at 72 mm. bore and 100 mm. stroke (3,255 c.c.).

The straight eight unit was a highly refined twin overhead camshaft design embodying the traditional Bugatti feature of a combined cylinder block and head arrangement, thus allowing ample and symmetrical water passages without the protuberance of stud bosses inherent in a conventional engine. The problems associated with cylinder head gaskets were also eliminated. The penalty for all this good sense was a loss of accessibility to the valves; however, the exhaust valve life of these engines proved to be outstanding, due no doubt to the good cooling provided by the layout. The camshafts were driven by a train of gears at the rear of the engine, fibre pinions were interposed with steel. The problem with a fibre gear is that of eventual fatigue failure with no warning, because they remain silent until the end. Rolls-Royce themselves adopted this type on their early Bentley Mk VI and Silver Wraith engines, and experienced the same problem, which they overcame by substituting a gear manufactured from an aluminium-bronze casting which proved both silent and enormously long lasting. The Bugatti crankshaft ran in four main bearings, plus one outrigger behind the timing gear. The engine's overall dimensions were very large for a capacity of 3.3 litre with a length of 3 ft 8 in. It was a beautiful piece of engineering sculpture. The gearbox of the Type 57 broke new ground for Bugatti in that it was in unit with the engine, together with a single plate clutch in place of the previous ingenious multi-plate unit. Synchromesh was never incor-

Bugatti Type 57 engine, the first example built. The gear lever is shorter than standard while the long plunger-type grease cylinders for the water pump are missing. The frosting effect on the crankcase and gearbox was not carried out on production engines, although retained for the camshaft covers.

The exhaust offtake was taken from a rearward position after the first batch.

Bugatti 57 Stelvio drophead coupé, an early example (1934–1935), the work of Gangloff of Colmar to the design of Jean Bugatti. Rear spats were often omitted.

A Type 57 with post-war British coachwork.

porated in this or any other Bugatti production gearbox, although adoption of the Cotal electrically operated epicyclic unit was planned for the successor to the Type 57 when development was abandoned with the outbreak of war.

A chassis frame of great stiffness by the standards of 1934 and incorporating the traditional reversed quarter elliptic rear springs was evolved. The mechanical braking system was the most unsatisfactory feature of the Bugatti, although reasonably powerful and working in the usual beautifully finished large diameter drums. The method of operation on the front axle was such that a self servo effect took place upon brake application due to the axle tending to rotate away from the operating arm, thus putting the brakes on harder. A frightful judder would ensue if the brakes were applied hard after wear of the brake cam and bush had taken place. It was also essential to back off the leading shoe lining.

Adjustment of these brakes was also a time consuming process involving removal of the hubs.

The radiator was now a modern film type block equipped with thermostatically operated shutters, the work of the Chausson concern who were the Serck of France. These shutters were not particularly well made, certainly not up to the standard of those on the Bentley. Steering and roadholding of the French car were vastly superior to the English product, while it also enjoyed an edge over the latter on performance. The main electrical equipment on the Bugatti by both Bosch and Scintilla was of excellent quality, but the effect was spoilt by the use of rather crude switches and fuse equipment. The English car was definitely superior in this respect.

Bugatti possessed coachbuilding facilities at Molsheim and in addition made use of the local concern of Gangloff Frères at Colmar. All bodies from these two shops appear to have been styled by Jean Bugatti; they were all striking in appearance while the Atalante fixed head coupé was arguably the

The Earl of Cholmondeley, a lifelong bugatti enthusaist, ordered this 3½ litre Bentley with rather untidy coachwork by the Parisian coachbuilder Van Vooren. These bodies were normally to be found on the K.6 Hispano Suiza. The external door hinges indicate extreme simplicity of construction and weight was reduced to an absolute minimum on this car. Rolls-Royce were apparently impressed by the quality of construction and later awarded Van Vooren an order for bodies for the Corniche, cancelled with the outbreak of war in 1939.

The Bugattis of the Earl of Cholmondeley: (left) a 1939 Type 57C, Figoni and Falaschi coachwork; (right) a 1938 Type 57 Atalante, Bugatti coachwork.

most beautiful motor car ever built, representing the zenith of the combination of art and engineering.

The Galibier four-door saloon was probably the least attractive; first series cars of this type were of a pillarless design, then enjoying a vogue in view of the ease of ingress and egress, but only when both front and rear doors were open together. The resultant lack of rigidity caused trouble in service while the general standard of construction of this particular type was indifferent. The later Galibier style was well built but heavier and not particularly elegant.

A large number of Type 57 chassis passed through various European coachbuilders to be fitted with fixed and drophead coupé coachwork, often with striking results, while James Young, Van Den Plas, Gurney Nutting and Corsica built a few one-off bodies on chassis supplied by the London Bugatti depot.

It is interesting to study the development of the Type 57 Bugatti with the equivalent Bentley. While the car from Alsace benefited from substantial improvements over five years, the Bentley deteriorated, principally due to the burden of a steady increase in weight. The 1937 season witnessed an increase in the cubic capacity of the Bentley to $4\frac{1}{4}$ litres, thus improving the performance, but, because of the weight, only marginally. The new engine was rougher than the $3\frac{1}{2}$ litre and had a short life if driven hard. Despite much thought and several changes of bearing metal specification, resulting in some improvement, a real cure proved elusive. The trouble was probably due to the thickness of the white metal big ends, poor heat transfer through the bearing shells and a general build up of crankcase heat. The sump capacity was modest for a pre-war engine of its size and the oil pump dimensions were surprisingly small.

Early $3\frac{1}{2}$ litre cars suffered from a disconcerting chassis tramp due to a lack of rigidity in the front of the frame. Salvation appeared in the form of the Wilmot Breeden stabiliser. This device consisted of a sprung front bumper with lead weights at the extremities which had the effect of damping out the troublesome oscillations. The majority of British makers were experiencing this problem in conjunction with the softening of front springs in order to improve the ride and Wilmot Breeden found eager buyers, Wolseley, Triumph, S.S. and Rover, to name but four.

The method of attachment to the front dumb irons was by way of flimsy U-bolts. Jack Barclay suffered a severe accident when the bumper of his demonstration car fell off and caused the car to overturn. The result was the addition of untidy claws around the spring blade in order to catch the bumper in the event of failure. The superior chassis design of the Type 57 Bugatti rendered heavy accessories of this kind superfluous.

Front springs and shackles on the Bentley were amazingly small resulting in poor life, before wear and excessive side float took place. Conventional bronze bushes were pressed into the Bentley spring eyes, which were easily replaced, whereas the Bugatti part was unbushed, thus requiring replacement of the entire main leaf when worn.

Substantial modifications were made to the Bentley for the 1939 season. The gearbox was redesigned to incorporate an overdrive top, 17 in. wheels were adopted, together with Marles steering gear and fixed radiator shutters in conjunction with a conventional thermostat. The car was marginally improved but the heavy feeling and mediocre handling remained. A standard Rolls-Royce pattern crankshaft damper was fitted to the Bentley engines and situated as it was, inside the timing cover, resulted in eventual seizure or break-up due to the effects of sludging, always most serious inside the timing chest. The present generation will not be aware of the appalling conditions that one found inside an engine before the advent of modern oils.

Bugatti also employed a friction crankshaft damper but it resided outside in the fresh air and thus did not suffer from this problem.

Coinciding with the arrival of the Bentley $4\frac{1}{4}$ litre came the Series II Bugatti Type 57. The principal change concerned the adoption of a flexibly mounted engine while a new chassis frame was evolved of outstanding stiffness, because the crankcase no longer aided in this respect. This change resulted in a car with almost turbine smoothness; the torsional problems associated with straight eight engines were now almost undetectable, whereas some roughness had intruded with the first series

A 1939 overdrive 4¼ litre steel saloon from Bentley. The elegance and lightness of the car built five years before has disappeared. The external door hinges are a surprisingly retrograde step, while the petrol filler is now external.

cars. The braking system was modified to eliminate the front self servo effect but the difficult adjustment remained. Total weight was kept down with the exception of one or two body styles, while the gradual addition of coachwork weight on the rear of the Bentley resulted in further deterioration of road holding qualities.

The year 1939 saw the revised Bentley mentioned previously, while hydraulic brakes were adopted for the Bugatti at the same time, thus eliminating the only serious shortcoming of this car. A supercharger had become optional by this time, a Roots instrument of typical Bugatti design which provided a mild boost, resulting in improved torque and better mixture distribution.

The weakness of the French franc provided a real advantage for the British buyer, and it is surprising that sales figures were so low. Orders in Europe had also become difficult, and were averaging two cars a week in the first quarter of 1939. The true worth of the Type 57 and 57C is reflected in 1984 values: examples command approximately six to eight times the price of an equivalent Bentley, although aided a little by comparative rarity.

A sports version of the Type 57 was offered from 1936 to 1938. This car differed substantially from the standard car and was probably the fastest production car in the world in pre-war days. No equivalent Bentley was produced. Evidence exists, however, of interest in a faster car, which can be seen in the Corniche prototypes, but the design project was unfortunately abandoned due to the outbreak of war.

The problems associated with a lack of capital at the Bugatti works and resultant severe limits on development and test work were to be seen in a survey of service problems. Forty-four per cent of the first 90 cars delivered between March and October 1934 were returned to the factory service department for rectification work as detailed below:

Fault	Number of cars
Engines damaged, principally seizures	15
Engine timing gear noise	21
Steering	9
Brakes	10
Bodies	10
Electrical equipment	5

36 engines repaired out of 90 cars delivered, i.e. 40 per cent

39 cars repaired (some with more than one fault) i.e. 44 per cent

The more relaxed atmosphere and protracted development and testing time at Rolls-Royce no doubt resulted in very low guarantee claims; however, precise figures seem difficult to obtain.

Another Jean Bugatti creation, exhibited at the 1936 Paris Show. The front wings turned with the wheels.

Bugatti 57S Atalante, 1936. A works photograph of an early 57S bearing the factory trade plate, outside the factory gates. The car is hiding the railway track which had recently been installed to expedite the delivery of railcars, then in production at Molsheim.

Bugatti 57S drophead coupe. A special sample body on the S chassis and the work of Gangloff to the design of Jean Bugatti.

Bugatti 57S Atalante, side view. Another early 57S Atalante – the door hinges were concealed on later cars. The body was produced in the Bugatti factory.

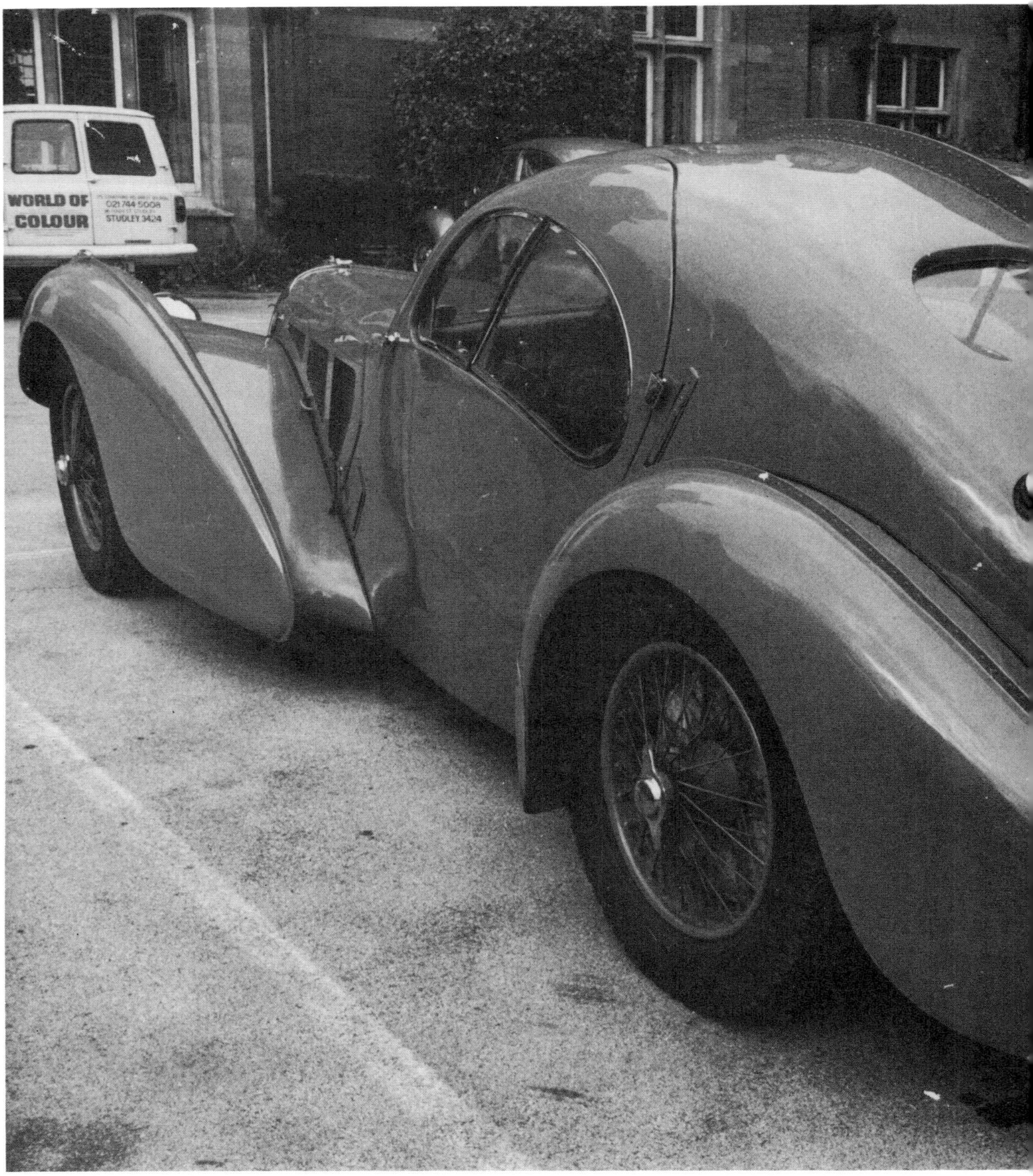
WORLD OF
COLOUR

A Bugatti publicity photograph of an Atalante coupé with roll-top roof on a normal Type 57 chassis.

The dramatic 57S Atlantique, designed by Jean Bugatti.

CHAPTER 8

1936–1939
The Phantom III, the 25/30 and Wraith

The advent of the twelve cylinder replacement for the Phantom II arrived in 1936. Despite a four-year gestation period the new car was plagued with serious problems, mainly concerned with cooling and oil filtration, the latter preventing satisfactory operation of the hydraulic tappets. The road performance of the Phantom III was, however, vastly better than previous models and Rolls-Royce now possessed a car capable of speeds which were the equal of offerings by other manufacturers and better than most. The ride and handling characteristics were also far better than formerly and with light coachwork was a pleasurable car to drive.

The new model possessed coil spring and wishbone independent front suspension and was given a chassis frame with good torsional rigidity to complement the softer springing; the arrangements at the rear remained conventional. A new gearbox with an even silkier action was evolved, although not as robust as previous designs. A reversion to earlier practice was seen in that the gearbox was separated from the engine and was placed well back in the frame; this position enabled a cruciform to be added to the chassis which crossed over ahead of the gearbox, thus adding stiffness to the most heavily loaded area of the frame, just aft of the engine bulkhead.

The engine was of similar dimensions to the 25 h.p. and $3\frac{1}{2}$ litre Bentley units: $3\frac{1}{4}$ in. bore × $4\frac{1}{2}$ in. stroke, and thus of a slightly smaller capacity than the previous big 6 cylinder model, at 7,338 c.c. The crankcase and cylinder heads were of aluminium, no doubt essential in order to keep the weight within reasonable bounds, but again a grave problem in later life due to corrosion problems.

The valve gear was operated by pushrods and rockers via the aforesaid hydraulic tappets, ultimately discarded in favour of conventional solid type. A dual choke downdraught Stromberg carburettor was specified, another example of the gradual switch to proprietary accessories. Water offtake rails ran along the top of each cylinder head and the engine was given a generally neat appearance.

The inherently perfect balance of the V12 layout coupled with meticulous Rolls-Royce attention in the matter of matching weights of reciprocating components resulted in the considerable performance being delivered with the utmost smoothness. The guarantee problems were, however, proving costly and embarrassing, and it is just possible that the makers may have been faintly relieved when the outbreak of war in September 1939 suspended further production.

It is difficult to find a European car with which to compare the Phantom III. With the imminent demise of Hispano Suiza, it is fair to say that Derby had the field to themselves from this point onwards.

A logical development of the 25 h.p. car appeared in 1936 when the 25/30 was announced. This car retained the 25 h.p. chassis, but performance was considerably improved by the fitting of a downdraught carburated $4\frac{1}{4}$ litre engine ($3\frac{1}{2}$ × $4\frac{1}{2}$ in. bore and stroke). Innovations were a Borg & Beck clutch and Marles cam and roller steering. The latter had in fact been specified for the final 25 h.p. cars.

A Phantom III prototype chassis, July 1935. Note the horizontal encased coil spring.

The 25/30 gained some 8–10 m.p.h. in maximum speed, and a light-bodied example would reach 80 m.p.h., with a corresponding increase in acceleration and low speed torque. The engine was, however, rougher than the 20/25, and suffered from a short life with regard to crankshaft bearings if driven hard. Total weight tended to go on climbing, which did little for the handling qualities. This car was superseded by the first Wraith in late 1938 which utilised a revamped engine in a modernised chassis with i.f.s. (independent front suspension), which was to continue until the outbreak of war.

The Wraiths gave a much improved ride, with markedly better handling than the previous cars. The much stiffer chassis allied with soft front spring ratings would be of great assistance to the traditional coachbuilder who, despite his craftmanship, was incapable of building an entirely satisfactory body on to a harshly sprung chassis, which itself possessed poor torsional rigidity. He was, in fact, expected to achieve the impossible, and this problem applied to all cars built in the vintage tradition.

A new gearbox was devised for this car and the lever was now fixed to the chassis, necessitating flexible mountings between this part and the gear

A Rolls-Royce Kestrel aero engine nearing completion. Note the valve operation by fingers between valve and cam as in the Bugatti Type 57.

selectors. This feature was not entirely satisfactory and the lever would fall out of engagement with the selectors on occasions. The life of the rear ball race also gave cause for concern.

Competition for the late '30s small Rolls-Royce had virtually disappeared, as it had for the larger cars. The small Hispano was listed until 1939, although production had virtually ceased by 1937, while the other French luxury makes were slanting towards more sporting carriages, although you could buy a Delage or Lago Talbot with formal coachwork. We can dismiss the clumsy teutonic offerings east of the Rhine. The Horch was possibly the most attractive of the German luxury cars at this period, but for sheer depressing ugliness the offerings of Maybach, despite excellent engineering, would be difficult to surpass. No Rolls-Royce equivalent could be found in Italy at this date, and one is forced to admit that it was to America that one looked for the ultimate in luxury. Vehicles which truly insulated their passengers from the outside world, and would waft them along at speed,

Prototype Phantom III chassis. Note the bolt-in cruciform.

and in utter silence, were to be found in profusion across the Atlantic.

The American market, to say nothing of their huge sales in export fields, was so immense that enormous sums could be found for development and testing. The final product would be made in such vast numbers that costs were quickly amortised. One needs to understand the entirely different emphasis which Americans were to place on aspects of motor car design, compared with European thinking. Comfort and smooth riding were paramount and, if need be, to the complete abandonment of road holding qualities, a fact brought home to the writer's father when, in 1938, he worked up a Lincoln-Zephyr to an indicated 95 m.p.h. between Leamington Spa and Kenilworth, only to find the greatest difficulty in controlling the car when confronted with the modest left hander at Chesford Brook. It was a fright from which he took some time to recover! The fact of the matter was, however, that the comfort and ride to be experienced in the cheapest Chevrolet was considerably superior to that of a Rolls-Royce or any other British or French production. The pressed

A main bearing from a 25/30 showing typical fatigue failure.

A 25/30 of 1938 bearing individual coachwork by Hooper.

steel body with which the Chevrolet was fitted was also a model in rigidity, silence, finish and long life, for these early steel bodies also appeared to have a much better resistance to corrosion.

The tendency to ostentation and vulgarity, with acres of chromium and other useless gimmicks, was not noticeably in evidence until the '40s, and in fact the appearance of some of the late '30s American cars, notably the Buick-Cadillac-La Salle line, were quite restrained and well balanced. There were, of course, exceptions: the Chrysler Airflow was unfortunate, although an excellent car, while the later Packards, apart from the Clipper series, became depressingly ugly.

The most notable of American cars of the '30s was probably the Cord 810–812 range. This machine, bristling with innovative features, was foisted on the public in an underveloped state, and the resultant guarantee problems presumably hastened the end of the Auburn-Cord-Dusenberg Corporation. The features of the Cord are too well known to merit repetition here, but if the front drive and transmission problems could have been surmounted, the result would have been formidable. The Cord possessed an extremely good performance delivered in the usual American manner of almost total silence; it gave an excellent ride, while the pressed steel body was finished to a very high standard. The attention to detail and quality of material to be found in the interior of the Beverley saloons, in particular, was quite outstanding, while the instrumentation was superior to that of any car in the writer's experience, and thus refreshingly different from other American cars.

Americans, obsessed with outside appearance, cared not one jot for underbonnet aesthetics. The sight of even quite expensive American engines was, therefore, dull, with little attempt at elegance or good finish. They were simply functional and the demand for maximum low speed torque, together with effortless high geared performance, led to an

1939 Wraith with normal production Hooper limousine body.

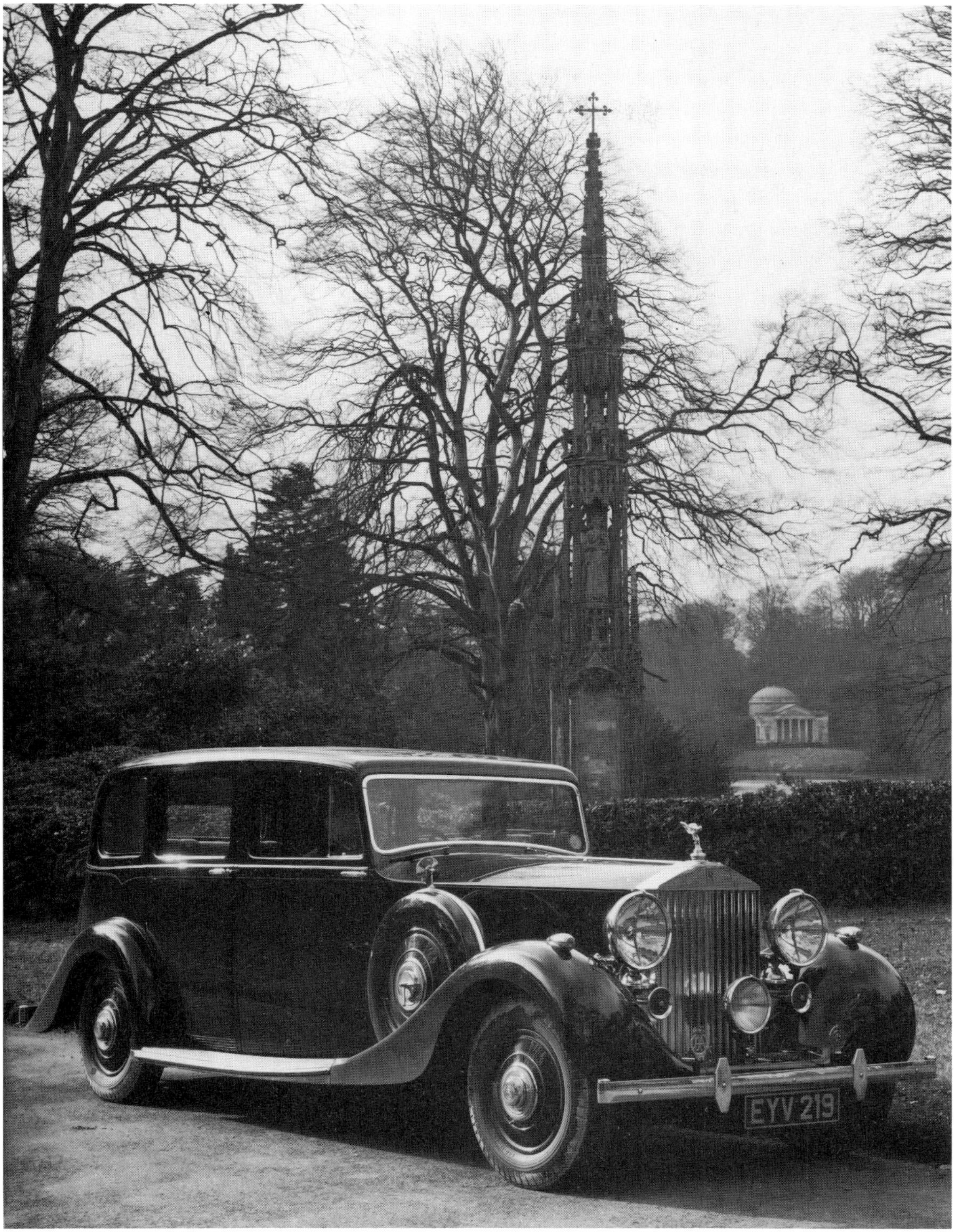
EYV 219

An experimental Straight 8 Bentley of 1939. A version of the engine in this car powered the very rare post-war Phantom IV.

Rolls-Royce were seriously considering an attack on the medium-sized car market in 1939. This short wheelbase car was built for appraisal. Note the odd pressed steel wheels. The mock up Park Ward body obviously formed a basis for the definitive post-war standard steel shape.

increase of cubic capacity in preference to a search for volumetric efficiency.

A large eight cylinder side valve unit was infinitely preferable to an efficient sports type engine of smaller size. The attendant anxieties caused by the raising of stresses with the latter type were certainly to be avoided. The writer well remembers a conversation which occurred between a representative of a firm trying to interest a certain large scale manufacturer in an efficient twin carburettor layout which had been developed for an engine and resulted in an increase in power of 8–10 h.p. The Chief Engineer replied that if they felt the need for another 8 h.p. they would add 2 mm. to the cylinder bores! An unusual feature for British buyers concerned the almost universal retention of 6 volt electrical systems by American manufacturers. The effort needed to start a large multi-cylindered engine in a North American winter was more than enough for the meagre system provided – perhaps all the owners possessed heated motor houses! – while the lighting offered was abysmal, a fact which underlined the excellence of American street-lighting systems. Other features which displeased the British buyer of these cars concerned long, whippy gear levers, tin dashboards faked to look like walnut, and cheap switches and controls.

The very large cubic capacity of all the trans-Atlantic top grade cars, renders a direct comparison with the small Rolls-Royce difficult; there was no equivalent, and in any case the aspirations of Derby and Detroit were totally different.

There was, however, plenty of collaboration between the British and American engineering staff. Packard, in particular, were to receive several deputations and the evidence of this liaison can be seen in the early i.f.s. layouts of Rolls-Royce cars.

It is interesting to imagine yourself as a potential buyer visiting the Motor Show held at Olympia in 1936. You might be prepared to part with £1,750–£1,800 for a 25/30 Rolls-Royce, although one could consider a 4½ litre Lagonda for around £1,100 or possibly even an Alvis Speed 25 for £800. You could, however, move further down market and contemplate the most expensive Buick Straight 8 for a mere £650; the £1,100 surplus from the Rolls-Royce price would allow you to buy a four-bedroomed house in Hampstead or half a dozen Austin 10s for your firm's commercial travellers. The Buick would not harm your reputation when parked outside your London club for the members would recognise it as identical to that which appeared almost daily in the press as the personal transport of Edward VIII. The car was considerably faster than the Rolls-Royce 25/30 and, furthermore, the performance was delivered in a manner most unobtrusive, while the effort required at the steering wheel was fractional compared with the English car, the ride was vastly superior, but perhaps excessively soft to the point where your children might suffer nausea. The commodious coachwork was stiffer, quieter and longer lasting than the efforts of the best London coachbuilders, while sealing from the ingress of dust and damp was infinitely better, windroar would be considerably less, and the Bedford cord upholstery would be cool in summer, warm in winter and kinder to one's suit. The main mechanical components would have an extremely long life, with the possible exception of the rear axle – certain Buick models were delicate in this respect. You would, however, have to tolerate cheap electrical equipment, a shoddy underbonnet appearance and poor exhaust system life, while the use of Mazak die castings for door handles and radiator grille would cause offence. You would not suffer too severely at the hands of the Buy British brigade, for the car was Empire built in Canada.

The Rootes Group, ever jealous of the success of the better General Motors products, were to field the Humber Snipe, a car aimed directly at the Buick market and about £100 cheaper in Britain, but probably dearer in export markets. It is fair to say that the Humber was inferior in every respect with the exception, once again, of electrical equipment and lighting.

CHAPTER 9

1946-1954
Post-War First Generation

The post-war motoring scene was instantly different from previous eras in one respect. The top grade luxury car market became the almost exclusive preserve of Rolls-Royce. A few concerns continued to build expensive sports cars; there were even new names in this field, Ferrari being the most notable, but the traditional Rolls-Royce competitors had disappeared from view.

Lagonda

Lagonda did not revive their expensive large cars, but attempted to market a 2½ litre sports saloon and coupé of advanced conception. This car remained in low volume production for several years, and survived one change of sponsor. It was, however, woefully undeveloped, and suffered from several serious shortcomings. It had all-round independent suspension by means of coil springs at the front and torsion bars at the rear. The car was distinctly unstable, and the rear wheels assumed the most peculiar angles. A number were written off as a result of serious accidents. An enlarged engine of 2.9 litres was adopted in 1954; although more powerful, it was rougher and had a short crankshaft life if driven hard. The ash-framed bodies, beautifully finished, were built by the manufacturers. A change in body styling was also effected in 1954; both styles suffered from ill-thought-out front wing and cowl stiffness, with the result that structural cracks developed in service. It is a pity that this car was spoilt by haphazard design, for it did possess a certain charm and character. Had matters been put to right, the Bentley Mark VI may have faced a worthy rival. The Lagonda must have lost money for its manufacturers.

Daimler

Daimler produced two new large cars of 27 h.p. (six cylinders) and 36 h.p. (eight cylinders) and while they continued to supply the Royal Household for a few more years, the cars were dull and cumbersome beyond belief. The chassis layout of these machines incorporated Dubonnet coil spring i.f.s., a system tried out before the war on the 2½ litre model. The 36 h.p. was the last British straight 8 engine to be manufactured, eventually being phased out in 1954.

The Austin Sheerline and Princess

The Austin Motor Co. made a concerted bid at capturing the relatively affluent with two new cars of 4 litre capacity, and a surprisingly good performance for vehicles emanating from the famous Longbridge plant. The cheaper model, called Sheerline, utilised a pressed steel body by Fisher & Ludlow of razor edge styling, as the name implied. The cars were trimmed and painted at Longbridge to a very good standard utilising the best of materials. An additional model built on the same chassis was to carry the quality theme one step further by the fitting of a really outstanding coachbuilt body by Van Den Plas, this London firm having been purchased by Austin for the purpose.

Austin Sheerline chassis. The extremely robust build of these cars is portrayed here, the influence of pre-war Austin design philosophy which produced the old, heavy 12 and 20 still apparent. Certain small parts were in fact carried over from the previous generation, namely wheel nuts, hub details, pedal gear, etc. The electric motor for the built in Smiths jacking system can be seen attached to the offside chassis side member. The rear axle was supplied by Salisbury Transmissions, the only occasion when Longbridge used a bought-in final drive.

Austin Sheerline production, 1948. Six early cars passing down the Longbridge lines, probably the first batch. Sales ran at a highly satisfactory level for three seasons (1949–1951), then tailed off, largely due to competition from the new and exciting Jaguar Mk VII. The final Sheerlines (1953–1954) were assembled at the nearby Kings Norton factory centre.

Austin Sheerline interior. The majority were finished in beige hide piped in brown, with top quality West of England cloth for headlining and Wilton carpets. Door casings were finished in cheaper material, but well-executed walnut veneers were in evidence. An alternative finish was dark-grey hide piped in red, with the interior woodwork given an attractive matt finish. The interior work was very well conceived; a neat touch was the flush fitting roof visors which can just be seen, a detail later copied by Jaguar.

These cars, the first to bear the Princess name, were extremely elegant, and considerably faster than the Rolls-Royce Silver Wraith. They were somewhat heavy and felt it, while a feeling of refinement, apart from the bodywork, eluded them. Perhaps one shouldn't expect it, for the list prices (ex. tax) in 1948 were £999 for the Sheerline and £1,350 for the Princess. The Rolls-Royce Silver Wraith standard Park Ward saloon would cost a purchaser £3,590 (ex. tax).

The chassis layout of the Sheerline and Princess witnessed the first independent front suspension system to be seen on an Austin car embodying wishbones and coil springing; a similar layout in scaled down form appeared a few weeks later with the announcement of the highly successful A40,

Austin Princess. The first example built seen on the slopes of Saintbury Hill near Broadway during the summer of 1947. The car was probably temporarily based at nearby Bibsworth House, the home of Austin Chairman, Leonard Lord (later Lord Lambury). This particular Princess appeared with a one-piece bonnet half; production examples incorporated the Sheerline interior bonnet release with detachable side panels.

This car would be fitted with a $3\frac{1}{2}$ litre engine; production versions were enlarged to 4 litres. (The lorry engines remained at $3\frac{1}{2}$ litres). The wing line on this Van Den Plas body design is similar to that used with less success on a handful of bodies built on Mk VI Bentley chassis during 1947. It is surprising that the Van Den Plas Bentley was so ugly, whereas the original Austin version was a model of elegance.

Austin Sheerline limousine. The first example of the stretched limousine version of the Sheerline in course of preparation in about 1950. The flush-fitting head and side lamps which resulted in a lighter appearance were not adopted for production; no doubt the small numbers envisaged did not warrant expensive modifications to the front wing tooling.

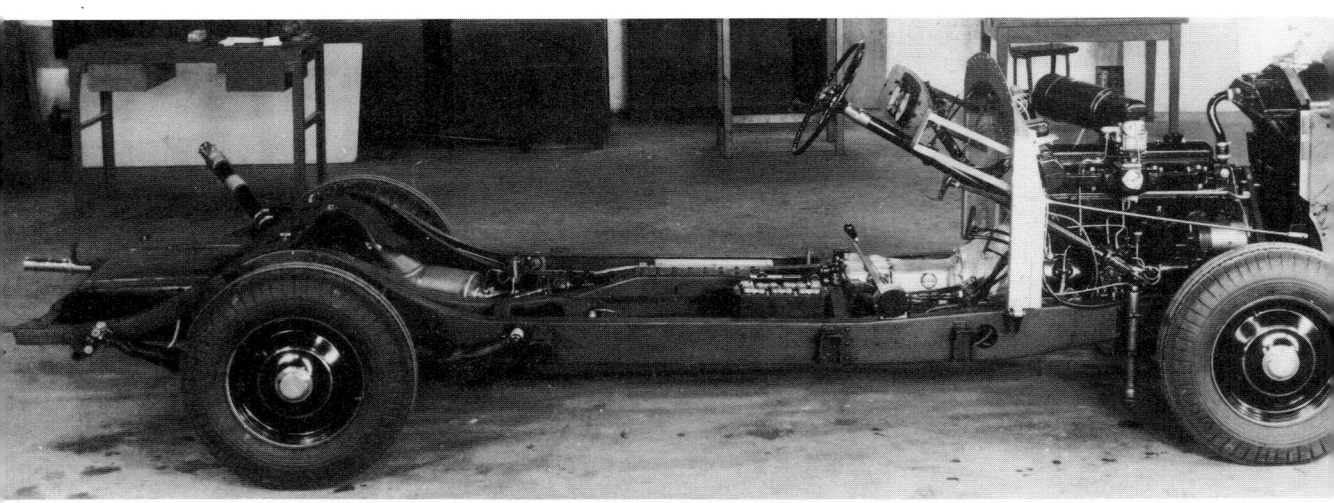

The first Silver Wraith chassis, in February 1946.

The first Silver Wraith i.f.s, rear view, with the Jackall system. This was abandoned for production cars.

which replaced the old 10 h.p. car first seen in 1932.

The engine of the large Austin bore a very strong resemblance to the truck unit first seen just before the war, and was none the worse for that; it was an excellent, smooth four bearing, six cylinder, pushrod operated, overhead valve design and was possessed of an enormous life. All the components of the chassis were extremely liberal in size, and an overhaul of any major unit was almost unknown. The gearbox, in particular, would have transmitted twice the power, and it is a pity that the Longbridge engineers made rather heavy weather of the steering column change mechanism, which was coarse in operation and inelegant in appearance. The car was spoilt in one or two other minor details: the exhaust system tended to have a very short life, while the spring steering wheel soon broke away around the spokes.

France

France tried to continue building some of their earlier models, and we saw Hotchkiss, Delage/Delahaye and Lago-Talbot cars of basic pre-war design until the early fifties, all very good cars, but doomed to fail in the difficult early post-war years. Harry Ainsworth, the English P.D.G. (M.D.) of Hotchkiss, said that they stood little chance of earning enough money to re-equip the works, their most modern machine tools had been taken to Germany during the war, never to return. Presumably this fate also affected the other makers.

Bugatti did not resume car production, although a few prototypes of various designs appeared spasmodically, while six to eight modernised versions of the pre-war type 57 were assembled, presumably to test the market. Hispano Suiza never reappeared.

Germany

The most serious opposition to the Best Car in the World was fielded by the Mercedes-Benz concern. There was clearly no place in the post-war world for their large and ostentatious supercharged cars of former days, so highly prized by the hierarchy of the Third Reich. In any case, they were disappointing in performance and had an outrageous thirst.

The first large Mercedes-Benz of the new generation was, in fact a 3 litre six cylinder car of traditional Stuttgart design and layout, utilising their tubular chassis and all independent suspension seen on previous models. This new machine was certainly refined, although somewhat lacking in performance. It found a ready market and a coupé version, known as the 300 S, was also offered, possibly the best looking car of its era. In later years we have seen the enormous and very ugly 600 limousine range. They lack the refinement of the earlier car, and appear inferior to the products of Crewe on almost all counts.

America

America has not entered the quality market since the war; the senior line Packards and Lincolns were the last survivors, but both were phased out by 1938.

The Bentley Mark VI

The first post-war generation of Rolls-Royce and Bentley cars were almost totally different from previous models; they were enormously simplified and much the better for that reason. The final pre-war Rolls-Royce, the Wraith, current for one season only, together with a handful of experimental Mk V Bentleys, indicated future thinking with regard to chassis and suspension layouts, but both of these cars still retained the old aluminium crankcase engines first seen in 1922.

The new Crewe-built Bentley and Rolls-Royce cars, which began to leave the works at the end of 1946, were basically identical in design, but the Rolls-Royce version, known as the Silver Wraith, was given a larger chassis in order to carry commodious coachwork, which would continue to be supplied by the London coachbuilders. The chassis frame of welded and riveted construction was stiff with an ample cruciform amidships, together with a large front cross member, while the side rails were boxed in forward of the cruciform. The rigidity achieved was better than that of any other British car of the period, but minor scuttle vibration problems did occur which resulted in late cars carrying additional channel section stiffeners

The first pressed steel Bentley Mk VI, April 1946, probably partly hand-built judging by the wavy panels. The hub caps were non-standard and rubber mudflaps were not yet fitted.

The first fully finished car but still without mudflaps.

MK VI Bentley, second stage. This car has the rearward facing mascot, waist strips, re-styled interior and export bumpers.

The final Mk VI, with side scuttle vents and new hub caps.

welded on to the outside of the chassis ahead of the bulkhead. Additional body mounts at the base of the scuttle were also found to assist. The new chassis was given independent front suspension with unequal length wishbones and coil springs, a development of that seen just prior to the war. This system gave good results and, although based on American designs, the roadholding achieved by Rolls-Royce was better than that accepted by the originators in the Oldsmobile-Buick and Packard camps. The standard of comfort was enormously improved while the life of the system, the achilles heel of many early i.f.s. layouts, was brought to an acceptable standard. The problems faced in service concern eventual wear of the bottom outer wishbone pivots, which incorporate needle roller bearings. The automatic chassis lubrication could and did fail and it is common to see the pin and rollers worn and rusted out beyond recognition in the case of a badly maintained car. The rust was caused by the ingress of water through the primitive (by modern standards) felt sealing rings. Wear in these pivots would upset the steering geometry by allowing negative camber; the knock-kneed appearance of a car in need of attention is instantly recognisable, while tyre wear problems also manifest themselves.

The king pins also ran in roller bearings, but fared rather better than the wishbone pivots unless the aforesaid one-shot lubrication system broke down. The upper outer wishbone pivot was by means of a rubber bonded Silentbloc bush. This feature was not totally satisfactory and the bush would eventually go eccentric, but worse, the strain of resisting rotation of the steering yoke under braking would allow the outer tube of the bush to move forward relative to the inner tube, until the yoke fouled the inner face of the wishbones. This state of affairs resulted in a loss of castor angle and would tend to make the otherwise light steering heavy in operation.

The lower wishbone was kept in place by a very stiff radius arm attached at its aft end by means of a large ball, complete with rubber cover, to a socket below the chassis frame. This radius arm was in the path of the split track rod and had, perforce, to be provided with a large hole through which the latter passed – a system identical to that seen on pre-war General Motors designs. It did of course impose a

Silver Wraith king pin failure, indicating the weakness at a change in diameter without a radius to avoid a sharp corner.

severe limitation on the allowable front wheel movement. The top wishbone was combined with a piston-type hydraulic shock absorber of typical Rolls-Royce design. This unit was excellent and considerably superior to the proprietary dampers then on the market. The split three-piece track rod necessarily involved relay levers which pivoted from the front chassis member. The resultant overhang would magnify the slightest wear or endplay in the pivots and result in some lost movement at the steering wheel. While this wear point would never reach a point of danger during the normal life of a car, it is a situation eagerly seized upon by over zealous M.O.T. mechanics who are inclined to fail a car even when the wear is at a comparatively early stage. The steering box, of worm and nut pattern, was once again of Rolls-Royce design and manufacture. It has a very long life and is beautifully smooth and light in operation.

American influence became apparent with the rear hub arrangement. This incorporated a ball race sealed for life and secured by a shrink fit collar on the axleshaft. This bearing can fail; the author had an embarrassing experience when one of these races broke up without warning and locked the rear axle solid on Magdalen Bridge, Oxford, during the morning rush hour traffic many years ago. The rear axle followed traditional Rolls-Royce practice and was very robust; an axleshaft or crown wheel and pinion failure is virtually unknown. The pinion race will eventually require replacement but warning is given with a gradual increase in noise. Unfortunately, the pinion bearing arrangement is special to Rolls-Royce and cannot be supplied by normal bearing factors. The set is very expensive.

The power unit employed in these cars is a fine piece of work. We now see a common cylinder block and crankcase of cast iron. The six cylinders have dimensions of $3\frac{1}{2}$ in. bore by $4\frac{1}{2}$ in. stroke, and thus similar to the previous pre-war model. The crankshaft, complete with bolt-on balance weights and fully machined, is supported in seven main bearings, while the crankcase extends $2\frac{1}{2}$ in. below the crank centre line. The engine is an admirably stiff design, aided by meticulous balancing resulting in exceeding smoothness.

The valve layout is of overhead inlet, side exhaust configuration, well known since the turn of the century, but largely abandoned by 1914. The principal British manufacturer to remain faithful to this layout was Humber, who adopted such a scheme circa 1923 following an investigation of the American Essex; all Humber engines were so built until late 1932. The old Bentley Company followed suit for their final model, the 4 litre, in this case, the work of Sir Harry Ricardo. Although somewhat expensive to produce, such an engine does offer freedom to incorporate a large inlet valve and port while the incoming charge lands on top of the exhaust valve offering a useful cooling advantage. The combustion chamber shape also allows good turbulence and adds to engine smoothness. Rolls-Royce continued with this engine range until the advent of the aluminium V8 unit in late 1959. The Rover Company brought out new engines of this configuration for their first generation of post-war cars in 1948. They were to continue until 1966 for private cars and for many years afterwards as an option for the Land Rover in six cylinder form. These power plants embodied a weird set of angles for the valves and cylinder head/block face.

The original $4\frac{1}{4}$ litre Rolls-Royce/Bentley engine continued until mid-1951 utilising a bypass oil filtration system which was discovered to be of very little use. The result of this scheme showed up in severe main bearing wear. The engines were, however, so well designed that they remained amazingly smooth, even when the crankshaft was badly scored with bearings worn right through the thin-wall white metal lining. The $4\frac{1}{2}$ litre engine which followed ($3\frac{5}{8}$ in. bore) was provided with a full flow oil filter which effected a complete cure; many of the previous engines were modified retrospectively.

Other modifications to the $4\frac{1}{4}$ litre engines included a change to aluminium bronze camshaft timing gears, as previously described. An increase in cylinder head stud diameter took place while clutch plate diameter was also increased. The throttle linkage layouts were also the subject of design improvements on the Bentley engines. The $4\frac{1}{2}$ litre unit eventually received larger big end bolts and an automatic choke was provided for the R type cars (September 1952 on).

The Rolls-Royce Silver Wraith and Silver Dawn utilised the same power plant, although de-rated by way of a single downdraught Stromberg carburettor in place of the twin S.U. instruments of the Bentley. These engines are comparatively simple to overhaul, utilise B.S.F. (British Standard Fine) threads and require very few special tools.

Problems in service are mainly concerned with the following points:

1 A short chromium liner was pressed into the top of the cylinder bore, no doubt a result of wear problems in experimental engines. This messy cure was worse than the disease. A ridge would form where the chromium met the iron, causing a break-up of the compression rings, while the resultant jolt which the piston received as it passed the ridge magnified a rattle which emanated from the small end, a noise particularly noticeable at tickover speed. The small end bushes must be in really perfect condition if silence is to be achieved.

2 The standard Rolls-Royce crankshaft damper, once again inside the timing cover, is subject to a break-up of the spring plate and suffers from heavy sludging. The original medium between the friction faces is by way of cotton duck washers, which needs laborious 'ironing' to bring the damper friction down to the designed figure. If the requisite frictional limits are not adhered to properly, the

Bentley Mk VI, 1949, chassis No. B-306–EY, prior to despatch to Trinidad after restoration by A. B. Price Ltd in 1980.

Accessibility on the first generation of post-war cars was good, aided in particular by an excellent quick release front wing and radiator assembly. The Silver Dawn this assembly belongs to, seen at the Hythe road service works, has now covered over 200,000 miles.

Bentley Mk VI, chassis No. B-56-AK. A typical case of heavy sludging to be found inside the crankshaft damper.

damper will not work and a noticeable period can be felt and heard at approximately 55 m.p.h., together with a lesser roughness at about 32 m.p.h. Fortunately, perhaps, the cotton washers are no longer available and the makers now supply these parts in a form of hard mica which cannot benefit from the ironing process; they seem satisfactory in service.

3 The exhaust valve guides of phosphor bronze appear to be subject to fairly rapid wear. It would be interesting to try cast iron guides but, of course, one hesitates to interfere with the manufacturers' specifications.

4 The removal of the aluminium cylinder head will be a difficult operation if it has been undisturbed for several years. This is due to corrosion around the cylinder head studs. There is no course other than to resort to the use of a specially designed and strong puller. Early engines with the small diameter studs appear to be the worst offenders and it is a good ploy to bore out the bosses in the cylinder head by an additional $\frac{1}{32}$ in., which will alleviate a future problem.

A new gearbox was evolved for the post-war car and a standard of silky and precise gearchanging was reached which has seldom been surpassed. Right hand drive cars were fitted with a well placed lever to the right of the driver's seat, while left hand drive cars were provided with a steering column change, fashionable at the time, and again of unsurpassed precision, all the more remarkable with a four-speed gearbox. A few cars were given a central gear change which was not a particularly tidy affair and lacked the feel of the standard pattern. This was evolved in the Crewe service department. The new gearbox, while basically of robust construction, did tend to suffer from a break up of parts of the synchromesh mechanism, principally concerned with the failure of fragile-looking 'ears' on the synchro cones, while parts of the detent ball retaining arrangements also gave trouble. Modifications were carried out during the production run and worthwhile improvements resulted. Chipping of the layshaft bottom gear will eventually occur, resulting in noisy bottom and reverse ratios. This problem appears to be beyond solution at the present time (1984). New laygear clusters have been unobtainable for several years and it appears that further manufacture is not contemplated. It would be certainly very costly to reproduce, for the layshaft is of one piece construction embodying four gears of helical pattern.

The condition of the brake servo operating cross shaft gear should be checked after a high mileage has been covered. These components enjoy a remarkably long life, but eventual wear will lead to failure and result in a total loss of brakes.

Exhaust systems on $4\frac{1}{4}$ litre cars were comparatively straightforward and of relatively long life. Pipes and silencers are connected by four bolt flanges utilising typical Rolls-Royce square-headed bolts. These should be retained if possible, because the bolt heads are designed to lock against the flanges, and so only one spanner is required to assemble. Unfortunately, they are nowadays fairly expensive. The $4\frac{1}{2}$ litre Mk VI was given a dual exhaust system of some complexity which wove in and out of the chassis on its way to the rear of the car. This system consisted of many short pipes together with four silencers. It is essential to align the system carefully to prevent fouling of the chassis and consequent noise and drumming. The R type Bentley system was simplified to a degree, while the equivalent Rolls-Royce retained the single system. The cost of replacement with maker's components is now absurdly high, a situation which has given rise to the manufacture of these parts by outside concerns, some of which are of very good quality and considerably cheaper than the original equipment.

The braking system adopted for these cars is unusual in that it represents a combination of parts by Lockheed, Girling and Rolls-Royce themselves. Mechanical brakes are awkward to use in conjunction with independent suspension and a Lockheed hydraulic set was provided for the front, but with adjustment by means of the standard Girling type expanding-shoe pivot. The rear brakes were of standard expanding wedge Girling pattern; the system was completed with the traditional Rolls-Royce mechanically driven servo motor. These brakes are excellent in operation and simple to maintain and adjust, an enormous improvement on the pre-war scheme. The condition of the servo liner is critical; both oily finger marks and hard glazing of the surface will render the servo inoperative.

Aluminium petrol tanks were introduced on the R type, resulting in a very small saving in weight in exchange for a serious service problem later. These tanks soon suffered from electrolytic corrosion, resulting in leaks. Canvas strips were then interposed between the steel holding straps. However, the position of the tank beneath the car, exposed to rain and damp, rendered the corrosion problem incapable of solution. S series cars were even worse and aluminium tanks were abandoned upon the introduction of the Silver Shadow in 1966, a return to manufacture by conventional tinned steel sheet being specified.

Rolls-Royce's painstaking and precise detail engineering can be seen in the pedal gear and related details. Great care has always been taken to eliminate draughts and fumes from entering the car via the pedal apertures. Neat little boxes containing felt ferrules are secured to the pedal apertures in the floor. This good sense is rather spoilt after a high mileage has been covered, for a minimum of wear on the pedal pivot will result in the pedal fouling the ferrule boxes, and a loud high frequency rattle will result.

The standard Bentley coachwork broke new ground for its makers, consisting of a four door, four light saloon shell provided by the Pressed Steel Co. and built to their all-welded and accurately-jigged formula. They were trimmed and painted at Crewe. Appearance bore a strong resemblance to the 1939–1940 Park Ward Wraith and experimental Bentley Mk V, although tidied up and improved in proportions. It was, in fact, a very good-looking car with well-engineered details. Door and window lift handles were, for instance, secured by large screwed rings requiring the use of a C spanner to fit or remove. Such parts were vastly superior to the fittings used by the majority of the independent coachbuilders, while rigidity, strength and sealing from wet and dust were all far better on the standard steel car. The door hinges were entirely concealed and of an unusual sliding construction. The Achilles heel of the standard steel body proved to be corrosion and they became chronic sufferers from this disease. The worst area concerns the rear wheel arches and wings, rear body mountings, sills and sundry other places where exposed to the ingress of mud and damp. The front wings were given a spot welded nacelle to house the side lamp and this area proved troublesome very early in life. Repair and restoration of a severe case can quite often result in expenditure in excess of £12,000. The more popular coach builders to whom Rolls-Royce supplied chassis were as follows:

1 Hooper & Co., responsible for a considerable quantity of Silver Wraith limousine bodies but only a few sports saloons on Bentley chassis, generally of striking, sometimes bizarre appearance. The quality of construction and standard of detail in bodies by this concern were superior by far to the work of their competitors. The business had passed into the hands of the B.S.A. Company and was thus engaged in the manufacture of bespoke coachwork for their subsidiary, the Daimler Co. Some commercial bodybuilding also took place. Ambulances were a speciality.

2 H.J. Mulliner & Co. Ltd. The main line for this company consisted of razor edge sports saloons of well-balanced appearance on the Bentley chassis. Their construction embodied the floor and scuttle of the standard steel saloon and, in early production, the boot and spare wheel door. Aluminium extruded sections were also freely used in the body framework, a medium also used by Hooper. The Mulliner concern seemed to attach more importance to the limitation of weight than their competitors and went so far as to call one of their later styles the Lightweight, although no material reduction took place. Silver Wraith touring saloon and limousine bodies were also built which bore a strong family resemblance to the smaller Bentley pattern. This old established concern, now a subsidiary of Rolls-Royce, built a very few coupés and virtually no work was undertaken for other makers, although two or three Humber Snipes appeared with Sedanca De Ville bodies identical to their first post-war offering on the Silver Wraith in 1947. Quality of construction and finish was always good.

3 The Park Ward concern, also now under the control of Rolls-Royce, specialised in drophead coupé coachwork for the Bentley chassis, although a considerable number of touring saloons were built on the Silver Wraith. These latter were positively

First post-war Mulliner Mk VI utilizing standard steel boot and spare wheel door.

First Park Ward drophead coupé on Mk VI chassis, 1946.

A youthful Harold Wilson with Sir Stafford Cripps at the British Exhibition in New York, October 1947, in a left-hand drive Bentley Mk VI drophead foursome coupé.

ugly, the first version consisting of an ill-proportioned four light design current up to 1949, followed with a long six light pattern with the front wing line flowing through both doors and embodying the now popular thin brass section window frames – chromium plated of course. It is fair to say that the construction of bodies in the post-war years by this previously respected maker were comparatively poor. Pressed steel was much used for inner door panels etc., and electrolytic corrosion took place where aluminium skins were attached to framework of a dissimilar metal, while door and window sealing were also poor.

4 Messrs Freestone & Webb were responsible for a smaller number of bodies on both Rolls-Royce and Bentley chassis, which generally followed the Mulliner style, but the fine proportions of the latter were lost and the method of construction utilised a large amount of timber in the pre-war manner. The final detail finish was, however, extremely good and the interior appointments were exquisitely carried out.

5 James Young of Bromley, a very old established firm which specialised in sports saloon and coupé coachwork in pre-war days and later under the control of Jack Barclay. Various designs of sports saloon bodies on the Bentley and Touring saloons for the Silver Wraith were offered together with a handful of coupés. It is fair to say that the work of this concern left much to be desired in the matter of sealing from draught and damp. In addition, their coachwork was structurally weak and susceptible to stress cracks in various places. Chronic electrolytic corrosion in the wings, sills and wheel arches soon appeared, and chromium plating appeared inferior to that of their competitors. They were also untidy in construction and the restoration of one of these

A small batch of Mk VI chassis received strange saloon bodies, the work of Van Den Plas. The wing line and rear spats appear almost identical to the Austin A135 Princess from the same house.

bodies will present a major problem to those concerned.

An automatic gearbox of General Motors hydramatic four-speed type was introduced as an alternative transmission during the production run of the R type. Towards the end of manufacture almost all cars were so fitted. The resultant two pedal control was at the expense of some performance and an increase in engine roughness, probably due to the loss of the heavy flywheel used in conjunction with the normal gearbox. Fuel consumption also suffered. These automatic gearboxes were robust in construction and virtually trouble-free in service. They were, however, highly critical with regard to throttle linkage angles and maladjustment would result in rough changes of ratio. Indeed, the tendency to jolt in the second to third gear upchange was never entirely eliminated and was probably the main reason for the eventual abandonment of this basic design early in the production life of the Silver Shadow. The Armstrong-Siddeley concern also used this gearbox in their $3\frac{1}{2}$ litre Sapphire model current at the same time, circa 1955. The torque characteristics of their smaller engine together with extremely flexible mountings exaggerated the jolt problem. The Parkside firm gave up the struggle and adopted a three-speed Borg Warner gearbox for their final Star Sapphire in 1958. Austin and Jensen also utilised the four-speed Rolls-Royce built box, which was quite successful with their larger engines.

The finest manifestation of this generation of Rolls-Royce products was undoubtedly the R type Continental, a high speed lightweight sports saloon of good aerodynamic form. This car, too well

H. J. Mulliner lightweight Bentley from 1951. A special two-door car built for Henry Ford II.

The first Bentley Continental prototype, with divided windscreen and lightweight bumpers.

Road dirt on the Continental provided an interesting study in aerodynamics.

known to describe in detail here, will rank as one of the most outstanding motor cars of all time. Full of character, it was capable of very high cruising speeds allied with good fuel economy, characteristics which were aided by a very high final drive ratio. Late examples were fitted with an engine enlarged to 4.9 litres by increasing the bore size to 3¾ in., while the automatic gearbox became an option towards the end of the model run, a feature which did not suit the personality of the Continental Bentley. The great majority of these cars were fitted with striking streamlined coachwork by H.J. Mulliner suited to the concept of the car. Several chassis were, however, supplied to other coachbuilders, with less than satisfactory results. The Parisian firm of Franay built a few almost slavish copies of the Mulliner style, but the proportions were lost and they were badly executed.

Park Ward did rather better with drophead and fixed head coupés, which were forerunners of their standard offering on the succeeding S.I Continental. The author owned the solitary Farina-bodied car for a period, and while this car was striking in appearance and quite good structurally, the detail features were inferior to the Mulliner product. Flimsy window winding mechanisms which required about fifteen turns from top to bottom are recalled as one example.

The principal problem of the R type Continental concerned a tendency to clutch shudder on take off due to the high gearing. This problem was never solved and was, apparently, the main reason for the adoption of the automatic gearbox. The standard Mulliner body usually cracked above the centre pillars and to a lesser extent at the top of the windscreen pillars, while a tendency to leak water around the bottom corners of the screen was apparent, which would eventually ruin the veneers on the dashboard.

A 1953 production Continental.

A late car with heavy Wilmot-Breedon bumpers and chrome strips on the waist and bottom edge.

A Park Ward S.1 Continental wood model.

CHAPTER 10

1955–1965
Post-War Second Generation

The second generation of post-war Rolls-Royce and Bentley cars were ready for production in April 1955, and consisted of a revised six cylinder engine of similar basic design to the previous pattern, and retaining the large bore (3¾ in.) of the later R type Continental. The new cars were considerably larger 17 ft 6 in. overall length compared with the original

An early S.1 chassis, prior to the body mounting section. Note the temporary rubber block to hold down the suspension. A coachbuilt chassis can be seen behind, already fitted with steering column and bulkhead, the latter an integral part of the standard steel saloon.

A pre-production S series in course of construction.

Mk VI at 16 ft, while width also increased by some $3\frac{1}{2}$ in. Naturally, kerb weight also showed an increase at 4,368 lb. (Mk VI, 4,088 lb.)

A new chassis was evolved, fully boxed but of thinner ($\frac{1}{16}$ in.) steel compared with the $\frac{1}{8}$ in. thickness of the previous model. Stiffening strips of $1\frac{3}{8}$ in. by $\frac{1}{16}$ in. were, however, spot welded on to the inner, top and bottom faces of the box. Rivets had finally disappeared from frame construction, which was now entirely welded. It was the work of the Thompson Group of Wolverhampton. The front suspension, although still of coil spring and wishbone layout, was entirely re-designed. Wishbone pivots now consisted of threaded knuckles which were bolted to pressed steel channel section arms. The upper pivot was combined with the shock absorber as in former types, although the damper was re-designed and enlarged in capacity. The king pin layout remained more or less as before, although the steering box was completely redesigned yet still of worm and roller layout.

The rear axle and suspension remained conventional and followed the previous model in scaled up form. A Z bar was now provided to restrain sway, while the oil operated ride control for the rear

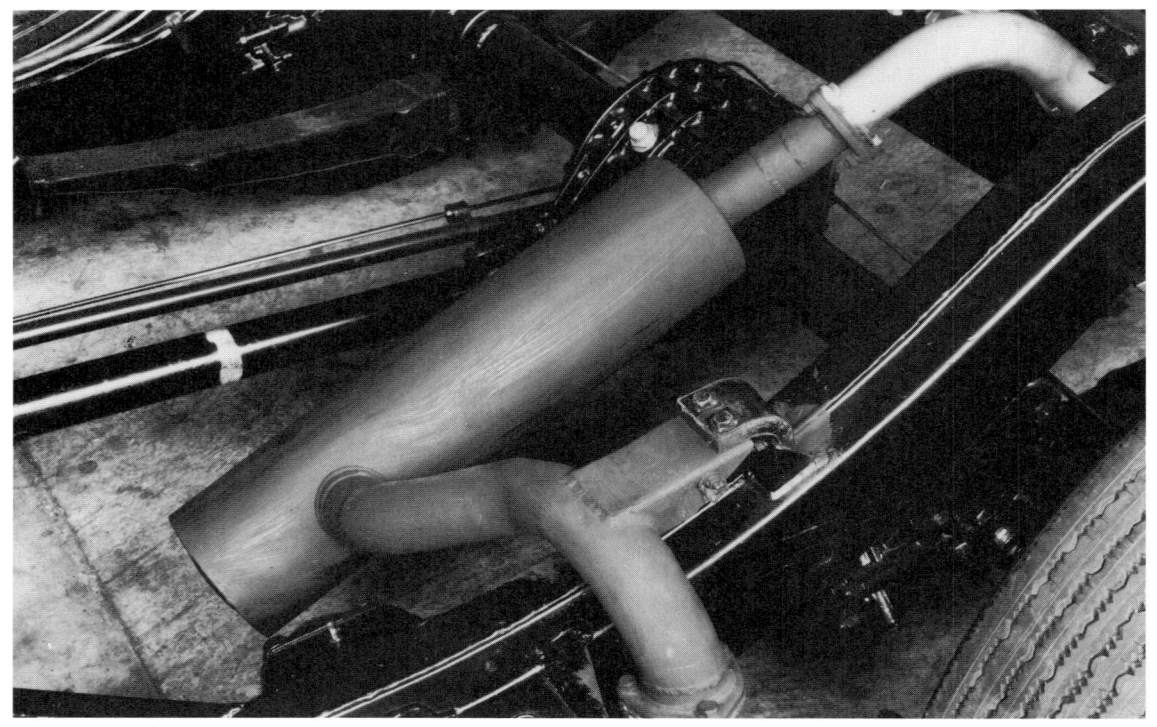

The complicated exhaust run on S series cars. The curved flexible mounting in shear was not particularly successful, while the exhaust pipe run over the chassis necessitated the use of heat resisting panels underneath the rear seat pan. This chassis was probably an early prototype since the anit-roll Z bar, fitting in production cars, is not in evidence.

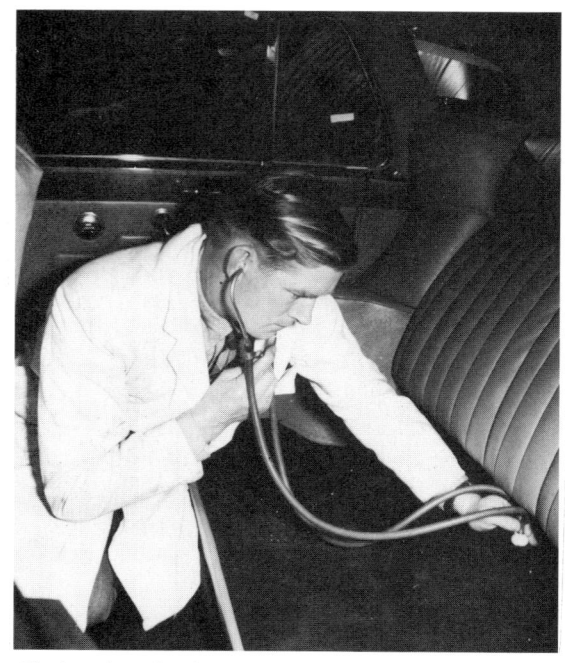

Checking for axle whine in a Silver Cloud II during the new car test curriculum.

The aluminium door construction does not appreciate the side impact test.

S.1 front suspension. Note the pressed steel wishbones.

S.2 front suspension, with forged wishbones.

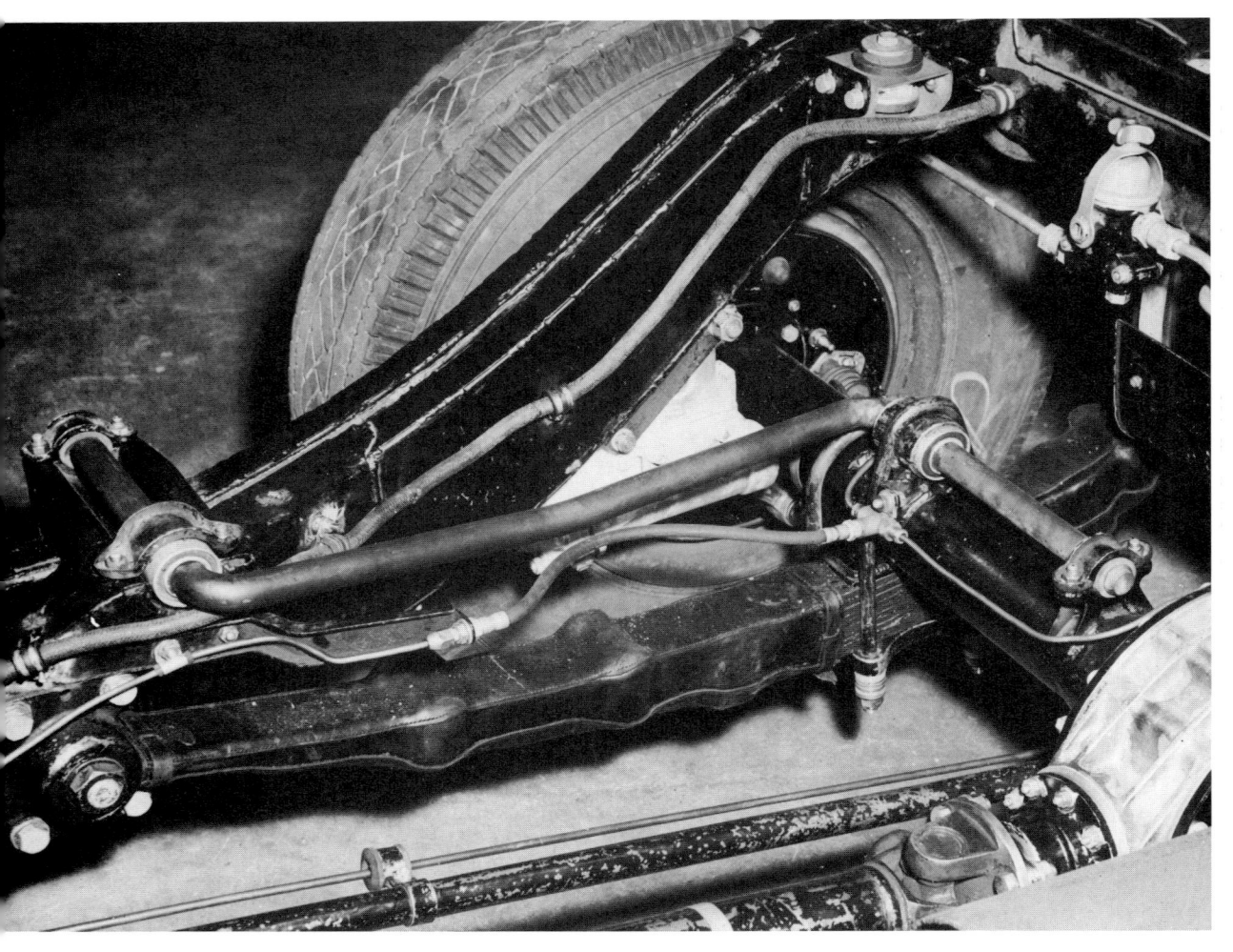

S.1 rear suspension, with anti-roll bar fitted to the offside.

dampers gave way to an electric solenoid pattern which was much simpler and eradicated a source of leaks. The braking system was now entirely hydraulic, aided once again by the faithful mechanical servo; however, the servo operating shaft was geared up in order to reduce the notorious lag in application at crawling speeds, which could be disconcerting.

The engine remained largely unchanged, although a new cylinder head was evolved which embodied half of the inlet manifold, this latter being split along its vertical centre line. A power increase resulted from the better breathing arrangements.

The head was secured to the block by bolts instead of studs and nuts, thus obviating the previous removal problems due to corrosion around the studs; the action of unscrewing the bolts, would, of course, break the bond caused by corrosion. The short chromium liners were also deleted. A.F. (American Fine) threads were now standardised.

The S.I engine was to prove extremely durable and it is common to discover units that have covered 200,000 miles without a major overhaul, and even then the crankshaft will probably be still within maker's limits of size. The hydramatic gearbox of the previous car was retained, although with additional clutch plates to retain an acceptable life, taking account of the increased weight and performance. The exhaust system reverted to a

The Phantom V test body used on all chassis. It was rapidly detachable and had lifting hooks. The wind roar, rattles and creaks from this mock up would seem to make such testing of little value.

The S.1 Continental chassis.

An S series wishbone pivot showing a worn out threaded knuckle – the thread on one side has totally disappeared.

single system embodying two silencers, together with a small expansion box in the tailpipe.

An entirely new and modernised four-door saloon body was evolved which embodied trailing doors, thin stainless steel window frames, a curved windscreen and a high wing line which flowed into the doors. The new coachwork was outstandingly well balanced and resulted in a very good looking car. Indeed, it may never be surpassed as an example of modern sculpture as applied to large saloon motor cars.

The interior trim design was also particularly neat, although the bench front seat, fitted as standard, was not satisfactory and gave no lateral support; retractable armrests were added for the second year of production. The S series car was undoubtedly quieter than the previous model. A big reduction in wind roar around the front pillars was achieved and insulation from vibrations and mechanical noise was of a very high order. Great care was given to body mountings and in particular the loading on each support was carefully controlled. To achieve this good result, an air operated device which resembled a milking machine was evolved for use when mounting the bodies to assist in obtaining equal loading at all points.

The new car was somewhat faster than the R type and still maintained a reasonably good fuel consumption. The least satisfactory feature concerned, once again, steering and road holding; it was heavy and lifeless, while the heavy tail made itself uncomfortably obvious in wet weather conditions. When front tyre wear took place the directional stability deteriorated and the heaviness increased still further. It is amazing that such mediocrity should have been allowed to mar an expensive car bearing such famous names. Furthermore, the life of the wishbone pivots was unsatisfactory, particularly the upper outer pivot, while the system, composed as it was of many parts bolted together, lacked rigidity. The one-shot oiling system was never entirely effective and was, in fact, soon removed from the track rod ends from where the majority of the oil poured.

Power-assisted steering was adopted during 1957, although at least one prototype was so fitted at least two years previously. This feature made for a vast improvement and eradicated the effort required to steer the car, while the feel appeared to be more precise than hitherto. Another facet which was troublesome concerned the heating and demisting system, complicated and operated by vacuum; the scheme was unreliable and not good enough even at the best of times. Modifications ensued during production which resulted in some improvement.

The most important change during the model run occurred in late 1957, when the engine received enlarged carburettors and enormous inlet valves; a further improvement in performance was the result. The rear axle proved a little delicate and an axle shaft would occassionally fail at the differential end, most unusual in modern times, while noise would develop after fairly high mileages and the eventual failure of the pinion teeth would ensue. The pinion bearings are too close together and the pinion lacks sufficient rigid support. The additional outrigger roller bearing on the nose of the pinion, a traditional Rolls-Royce feature, is too small, and is therefore overloaded, a state of affairs which can be verified by inspection of the inner track on a high mileage car. A stiffer axle was provided for the last year of production. The final cars were delivered in August 1959.

Later in life it became apparent that the S series was to suffer from corrosion problems on a scale even worse than the Mk VI. The rear wing inner and outer skins were the first to suffer, together with the trailing edge of the front wings where spot-welded to a baffle. Further trouble appeared above the front wheel centre line where wet mud would collect around a stay. The forward end of the body sills were provided with a welded in diaphragm to seal the box and this part would give way, thus exposing the entire sill to the ingress of mud and dirt while the adjacent body mounting would virtually disappear. The chassis frame also eventually gave trouble in the vicinity of the battery at the offside rear extremity. The body mounting outriggers at the front would also suffer and, in extreme cases, the main side rails where they arched over the rear axle would rust away. An enthusiast contemplating the purchase of one of these cars would do well to remember the phrase *caveat emptor*.

The range of coachbuilt bodies for the S series chassis was considerably reduced and consisted of the following:

1 Park Ward. This concern did not build on the standard car but concentrated on the Continental chassis, for which both fixed and drophead coupé styles were provided. They were well proportioned and superior in construction to their previous offerings. The Willesden firm also offered a new style on the elongated Silver Wraith, which continued in production until 1958. These bodies, although heavy, were generally improved in the quality of their construction and appearance compared with previous offerings. They then undertook work on long wheelbase standard steel bodies.

2 Hooper. A new style was evolved for the S series built in the traditional Hooper manner, although some effort had been made to reduce weight. The appearance of these cars, with their weird cowled headlamps was something of an acquired taste. One or two bodies of strange appearance appeared on

The stretched version of the standard steel saloon. There is ample lead loading in evidence.

An S.1 Continental outside Kenilworth Castle. The original very good proportions are now lost.

Continental chassis in 1958–1959. A further effort to reduce weight was made to a point where the rigidity and life of the body was seriously affected. They were also dreadfully ugly. The traditional razor edge style continued on the Silver Wraith line until the end, and almost all retained external P.100 headlamps. The latter were notable for providing minimum candle power for maximum wind resistance.

3 Freestone & Webb. The longer chassis of the S type car provided the other Willesden firm with an opportunity to extend their swept razor edge styling. The result was tremendously elegant, if a trifle ostentatious. Two flamboyant drophead coupés were also built before the closure of this old established firm in 1959.

4 H.J. Mulliner. A few bodies of cumbersome

appearance were constructed on the standard chassis, but work concentrated on the Continental. The streamlined R type Continental was enlarged and altered for the new model. The fine proportions were, however, lost to a degree. This style gave way to a notched back for the 1959 season. A much more successful option appeared in 1957 in the form of a four door, six light saloon known as the Flying Spur. This model proved to be the best of all the S series Continentals and the type was to continue for nine years virtually unchanged; a very few four light versions were also executed. They were all to prove corrosion resistant and good sealing was achieved.

5 James Young. The increased rigidity of the S series chassis may have proved helpful, but the new range of bodies provided by the Bromley firm were much better than former types and in addition were extremely elegant. The provision of adequate sealing, particularly around the back light was, however, still proving difficult, a problem compounded with the enlarged wraparound styling now fashionable. The quality of chromium plate still remained poor, while the seating provided appeared mean compared with that of their competitors.

A rare four-light Flying Spur. The first example, built in 1956.

British craftsmanship at its best. The interior of the four-light Flying Spur.

The H. J. Mulliner Flying Spur Bentley Continental, the best of all the coachbuilt bodies. The 1938 Warwickshire number plate was often used on demonstration Continental models.

Bentley S.II 1959–1962/Rolls-Royce Silver Cloud II

It became known that Rolls-Royce were developing an aluminium V8 engine, and in fact prototype cars so fitted were certainly on the road in 1957. The new power plant was to be of 6,230 c.c. capacity with the over-square dimensions of 4.1 in. bore × 3.6 in. stroke. The S type car accepted this engine and deliveries commenced in August 1959. The basic design is still current (1984) and has been the subject of steady development in the Rolls-Royce manner. The use of aluminium for basic components such as cylinder blocks and heads is fraught with problems. The inherently greater rate of expansion caused by heat over that of cast iron is one factor, whilst a light alloy cylinder block requires the use of a separate cylinder liner. The latter gives rise to a potential snag in the form of water seepage into the sump. The Rolls-Royce engine has proved very successful from this point of view and it is fair to say that problems connected with cylinder sealing are rare. Three O rings are provided with a drain to atmosphere between the lower two. The cylinder block extends $3\frac{3}{4}$ in. below the crankshaft centre line and five main bearings of $2\frac{1}{2}$ in. diameter are provided, pin diameter being $2\frac{1}{4}$ in.

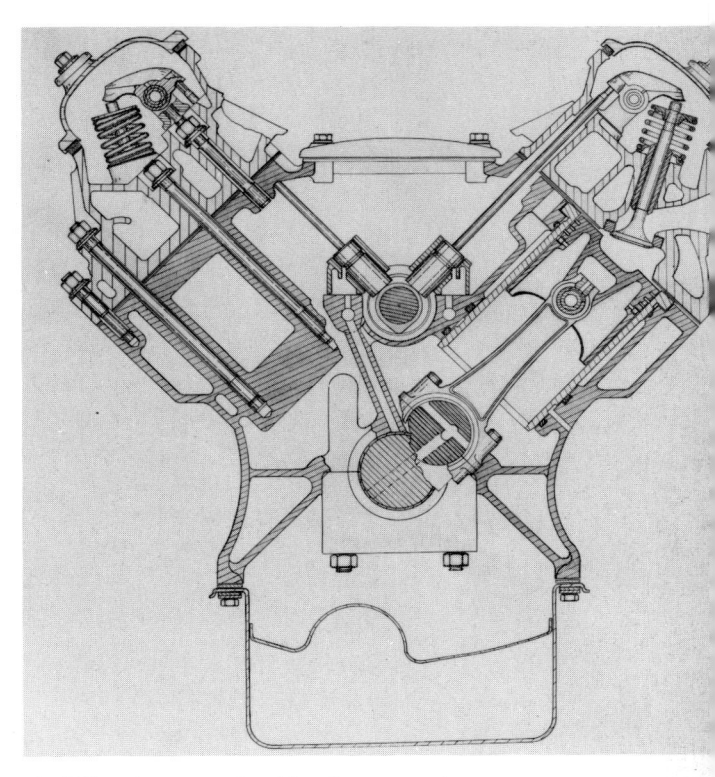

An S.II engine cross section. Note the concave piston crown. The higher compression S.III engine utilized a flat-top piston.

The use of bolt-on counter weights was now abandoned, while the old Rolls-Royce style of crankshaft damper disappeared at last, to be replaced by a simple bonded rubber Metalastic type which was fitted externally to the crank nose. The con rod was well proportioned, but the costly business of machining this component all over was now abandoned. A single central camshaft was driven by the usual helical gearing; hydraulic tappets, originally supplied by the Chrysler corporation, transmitted lift to the pushrod-operated valve gear, and both inlet and exhaust were now overhead and in line. Twin side draught S.U. carburettors were provided, which fed mixture to an octopus-shaped inlet manifold. The hydraulic tappets, aided by an efficient oil filtration system, now had a reasonable chance of surviving. An occasional tap from this source has to be tolerated when starting from cold, while the tappet base is subject to eventual wear or starring – a state of affairs which should not be allowed to continue, or damage to cam lobes will result. Shim steel cylinder head gaskets were to prove troublesome; a change to an asbestos woven pattern put matters to rights.

The S.II engine was prone to suffer from a carburation flat spot and was a little difficult to maintain in tune. The general performance was superior to the previous six cylinder unit, but not markedly so; in fact, when one considers the size of the engine, it could be considered disappointing. The valve rockers on the first years' production were fitted with a bronze bush and the rocker shafts were subject to very rapid wear, which manifested itself by a wheezing sound. A change to a chilled iron rocker running direct on the shaft cured this problem and most of these engines were converted during the guarantee period.

There were some failures of the oil pump drive on these V8 engines, which consists of a worm on the nose of the crankshaft meshing with a wheel on

An S.II engine. Note the very short, stiff rockers.

The S.II crankcase.

the oil pump drive shaft. Such drives are subject to a high pressure rubbing action, and the latter part will wear to a knife edge, and then fail completely. There is no warning, and unless the engine is stopped instantly a total ruination of the crankshaft, pistons, liners and bearings will result. The author has noted failures on cars after a mileage ranging from 65,000 to 125,000 miles. The differing length of life of this component, although presumably due to lubrication, is unexplained. It would be prudent for owners of high mileage cars to reassure themselves as to the condition of this part, which can be inspected after removal of the sump.

The front suspension was re-designed in detail and was vastly improved as a result. The wishbones were now one-piece rigid forgings, while the upper outer pivots were considerably enlarged and the one-shot oiling system was, at last, discarded. Lubrication was now by way of individual grease nipples of the hexagonal type found on most self-respecting vintage cars.

The width of the V8 engine necessitated a new steering box of great complexity placed outboard of the chassis frame. Effort was transmitted from the steering wheel by way of a pair of relay gears at the base of the column. A smaller and slimmer steering wheel was provided, which resulted in a small increase in effort on the part of the driver.

A scaled up rear axle was adopted and the final drive ratio rose from 3.42 to 1 in the case of the standard S.I to 3.08 to 1. Conversely, the Continental chassis gearing was lowered. These cars now utilised the standard crown wheel and pinion of 3.42 to 1 in place of the ultra-high 2.92 to 1 set fitted to the previous six cylinder model. The V8 axle was entirely trouble-free.

The remainder of the chassis remained as before, apart from a revised exhaust layout, while the automatic gearbox received yet more drive plates to cope with the increased torque. The throttle kick-down linkage was modified and made more awkward to service by virtue of adjustment from under the car on the side of the gearbox, instead of the accessible position above the starter motor on the six cylinder cars. The hydramatic gearbox was fully up to the task of transmitting the torque from $6\frac{1}{4}$ litres of engine and good life can be expected from the standard car. I have noted, however, that the much heavier Phantom V, a car by its very nature often committed to slow-speed city work, will need a replacement transmission twice as often as the standard car.

The S.II Continental was the fastest car produced by Rolls-Royce up to this period. The author purchased the prototype (Chassis No. 26B) from the works, after development had been completed. This machine was based on a 1955 Park Ward S.I coupé and was capable of exceeding 120 m.p.h. The braking system was obviously a cause for some concern, and a four-shoe scheme was adopted on the front for this model. The shoes were free to pivot on carrier shoes, and the object was to increase the lining area. This layout continued for the life of

A Bentley S.II chassis, showing the exhaust passing over the chassis.

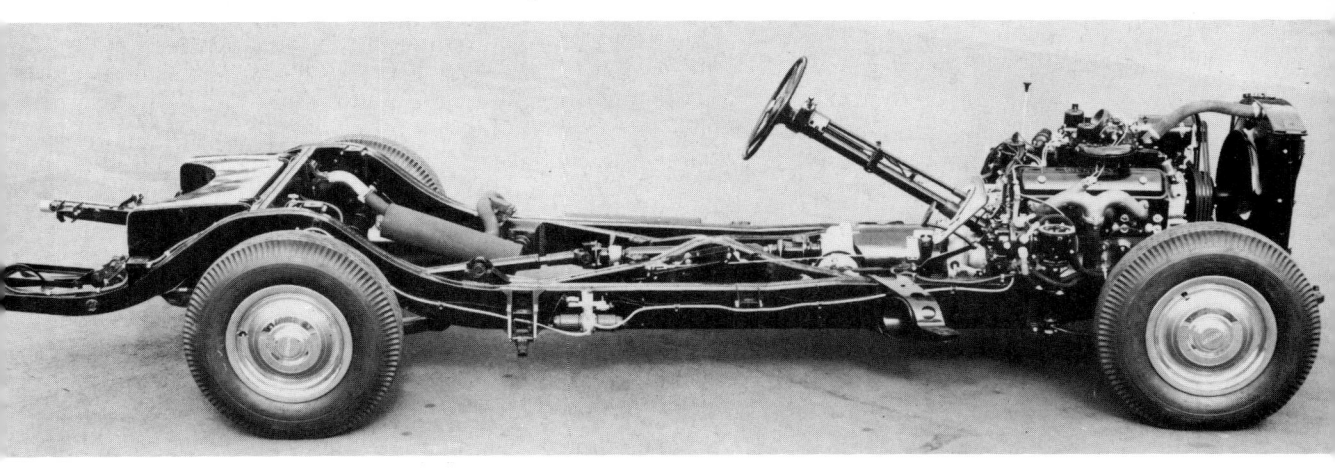

A Silver Cloud III chassis prepared for exhibition, with temporary brackets to carry the radiator, normally located by the wing valences. Note the repositioned exhaust below the chassis.

A Phantom V chassis.

A Bentley S.3 30 m.p.h. frontal impact test. A serious interest was being taken in deformation following accidents years before demand by legislation.

The Park Ward coupé evolved for the Continental. Inferior in construction and design to the Mulliner version and a chronic sufferer from corrosion.

the S.II, and for the first eighty-odd S.III Continental cars. It was then abandoned, as any beneficial effect was presumably found to be marginal; production then reverted to the original standard saloon two-shoe system.

The heating system was re-designed for the S.II and embodied electrically operated flaps in lieu of vacuum, while the ducting to the interior of the car was modernised. The new scheme was an undoubted improvement, but an acceptable standard of reliability still eluded its makers.

The Rolls-Royce car had now passed the point where even routine servicing could be carried out by an enthusiastic owner. The accessibility of sparking plugs was probably the worst ever achieved, with the possible exception of certain Ferrari models. This task necessitated jacking up the car and removing the front wheels together with access plates in the engine side valences! Furthermore, one was compelled to work in the dirtiest area of the car and mechanics had to endure a baptism of road dirt which would descend on them from the underside of the wings.

The coachbuilding trade had contracted further and only three firms offered alternatives to the standard car.

1 Park Wark did not offer bodies on the standard car, but evolved an entirely new Continental body which was offered in two door saloon and drophead coupe form. The two types were similar below the waistline, of inelegant, slab-sided appearance. They were to suffer from the most outrageous corrosion problems, which would affect the entire structure in the course of time. The construction was of steel, although boot lid, bonnet, door skins and front wings were of light alloy. Park Ward were also building bodies of vaguely similar styling on the 3

Bentley Continental body framework at Park Ward exposing the poor and untidy design. Note the different forms of welding utilized. Second series Phantom V body frames are in the background.

Alvis body construction at the Park Ward works, Willesden, 1959. First stage in jigging up a saloon, with the rear wheel arch and bulkhead fixture behind.

Constructional work nearing completion.

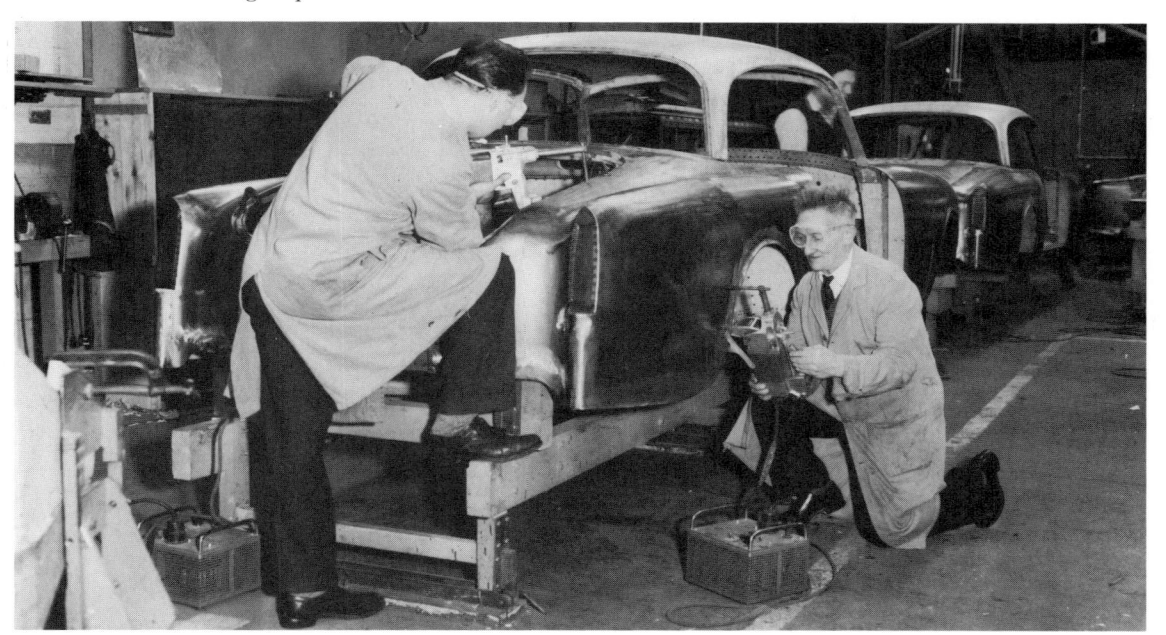

Saloons and coupés on the panelling line.

A wooden horse for shaping Phantom V wings and doors at the Mulliner/Park Ward works. The untidy shop indicates poor housekeeping.

Checking door clearances at Park Ward prior to passing for the paint shop; first series Phantom V

Prototype P.5 built in 1957 with a Hooper body. Purchased from Rolls-Royce by the author and later converted to the double headlamp system.

Bentley S.III Rolls-Royce Silver Cloud III

The final manifestation of the S range appeared in September 1962 and was readily identified by virtue of a lowered radiator and bonnet line, together with re-contoured front wings embodying a twin headlamp system.

The engine received high compression pistons and enlarged gudgeon pins, together with a modified induction system. The performance was improved, while the old tendency to suffer from a carburation flat spot disappeared. Gearbox kick down adjustment was simplified and steering castor angles were revised. The steering became lighter and more precise while the lowered bonnet line greatly aided visibility for the driver.

Separate front seats were provided which were vastly better, while room in the rear compartment was increased. Heater ducting was altered once more together with various other detail changes.

Modifications to the S.III during the three-year production run were few. The exhaust system was altered after the first series to incorporate bell-mouthed flanges secured by half-round clamps in place of the previous three-bolt pattern. The front seats received heavily bolstered cushion collars during 1964, rather spoiling the comfort provided previously, so far as the author is concerned. The last few cars can be recognised by one small detail: the roof lamp switch on the driver's centre pillar was placed horizontally instead of vertically.

Problems encountered in service were minor. An occasional car would develop a peculiar cam gear rattle, which was subdued by shrinking on an extra weight to the periphery of the damper. This modification would thus constrain a crank torsional vibration which caused the noise. The most serious matter concerned the incipient corrosion which would attack body and chassis. The front wings were now manufactured in two parts which were united by a roll weld just above a swage line. Rust appeared between the panels at this point within months of delivery. An abortive attempt to conceal this problem by the fitting of fussy chrome strips along the weld line was offered to complaining owners. The exhaust system life was also poor for a car of this calibre.

Rolls-Royce head office at Pyms Lane, Crewe, built in 1939. The car side of the business was run from here in the post-war years

The Mulliner coupé utilizing the standard steel saloon as a basis.

A Phantom VI chassis, with some Shadow features beginning to appear.

A projected replacement for the S.3 rigged up for a crash test. The silver Shadow shape has not quite yet crystallised and we see strong Austin Princess influences at the front, with a Hooper continental window line.

The coachbuilt line continued as before, although the twin headlamp system was adopted for all styles in line with that of the standard car. Messrs Park Ward and Mulliner, both subsidiaries of Rolls-Royce, were now trading as one unit. Bodies from both houses now bore the combined H.J.M.P.W. initials on heel mats and the like.

The S.III range, representing cars which had undergone continuous development over a period of seven to eight years, were very fine machines, despite the problems mentioned above. A better car in which to undertake a long continental journey is still hard to find. They provided a very high standard of silence, refinement and performance and were utterly superior to any other car in production at this time and, on many counts, to the early series of Silver Shadow cars which followed at the end of 1965. Their character can, perhaps, best be summed up by the expression of an old Rolls-Royce salesman when trying to conclude a sale to a gentleman with a member of the fair sex on his arm: 'It is a man's car, Sir.'

CHAPTER 11

Examples of Rolls-Royce Engineering Style

Connecting Rod Design

The most highly-stressed and central component of a piston engine is the connecting rod assembly. It is of interest, therefore, to study the evolution of this part during the first 60 years of motor car production by Rolls-Royce.

The original Silver Ghost short stroke engine utilised a well-shaped, but unmachined, steel forging; a half-hearted attempt to file down rough edges was made. The oil feed to the small end was by way of a straight copper pipe soldered into position, and clipped to the con rod by two simple pieces of bent wire. The big end bolt was of $\frac{7}{16}$ in. dia. B.S.F. thread (18 t.p.i. – threads per inch) waisted top and bottom and fitted with a simple slotted nut. Bronze big end shims of $\frac{1}{8}$ in. thickness were used. The bearing width is surprising and equals the diameter. Small ends were fully floating as in all Rolls-Royce engines.

The long stroke Silver Ghost engines were fitted with a fully machined connecting rod, and were shorter than the previous type, in order to retain the same basic engine height but to allow the use of a longer crank throw. The rod was very well proportioned and must have had a large margin of safety. The small end feed pipe was now neatly clipped and riveted to the rod. The clips were then soldered over for safety; these clips, however, tend to spoil the clean design of the rod and are always a possible source of trouble.

Then came the first appearance of the now traditional Rolls-Royce big end bolt, $\frac{7}{16}$ in. diameter, but with a special fine thread of 22 t.p.i. The bolt is no longer waisted but drilled from the top to a point just above the start of the thread. The nut incorporates a washer and the hexagon is reduced in size to that of a standard $\frac{3}{8}$ in. B.S.F. The crankshafts of these engines were considerably stiffened with thicker webs, resulting in narrower, but still ample, journal width. The bronze shims were now lined with white metal, a feature of doubtful merit which was continued until 1939. The metal was prone to chip and break off, with a resultant loss of oil pressure.

The stroke of the Phantom I overhead valve engine was further lengthened and the connecting rods were quite a bit longer and less robust than the Ghost, with a smaller big end diameter; the fine proportions of the previous design were lost. The small end oil feed pipe was now serpentine, no doubt to allow for temperature differences, to be taken care of by allowing the pipe to expand slightly if necessary; some fractures had probably occurred with the previous straight pipes.

The 20 h.p. car introduced in 1922 was provided with a smaller edition of the Silver Ghost connecting rod; the 1927 example in figure 156 (a) incorporates the serpentine small end feed pipe, and this configuration may have utilised a similar component. A diamond-shaped rail, fully machined, ran along the centre of the rod in order to make possible a central drilled oilway to feed the small end, thus obviating the untidy and potentially troublesome external pipe.

Figure 156 (c) shows the Mark VI and Silver

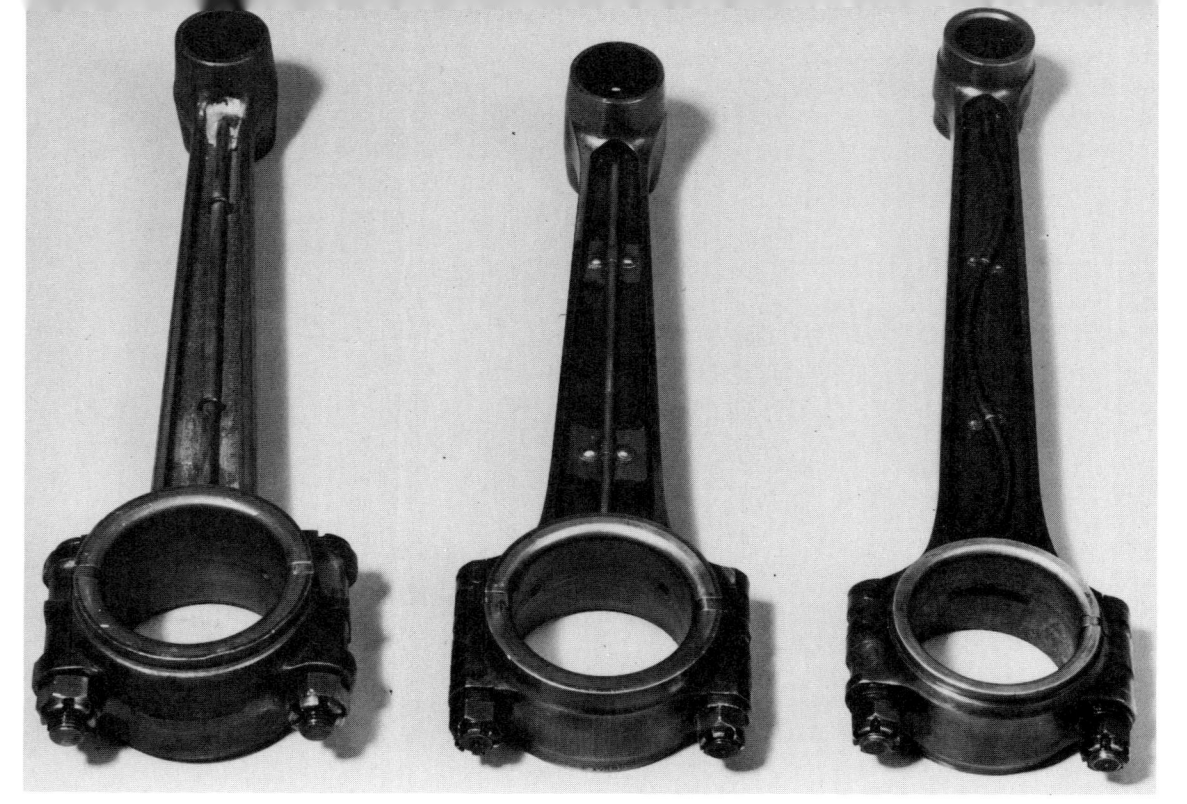

Rolls-Royce connecting rods. From left to right: *Silver Ghost 1907–1909, Silver Ghost 1910–1925, Phantom I.*

Rolls-Royce connecting rods. Left to right: *20 h.p., 25/30, Silver Wraith, S.II, S.III.*

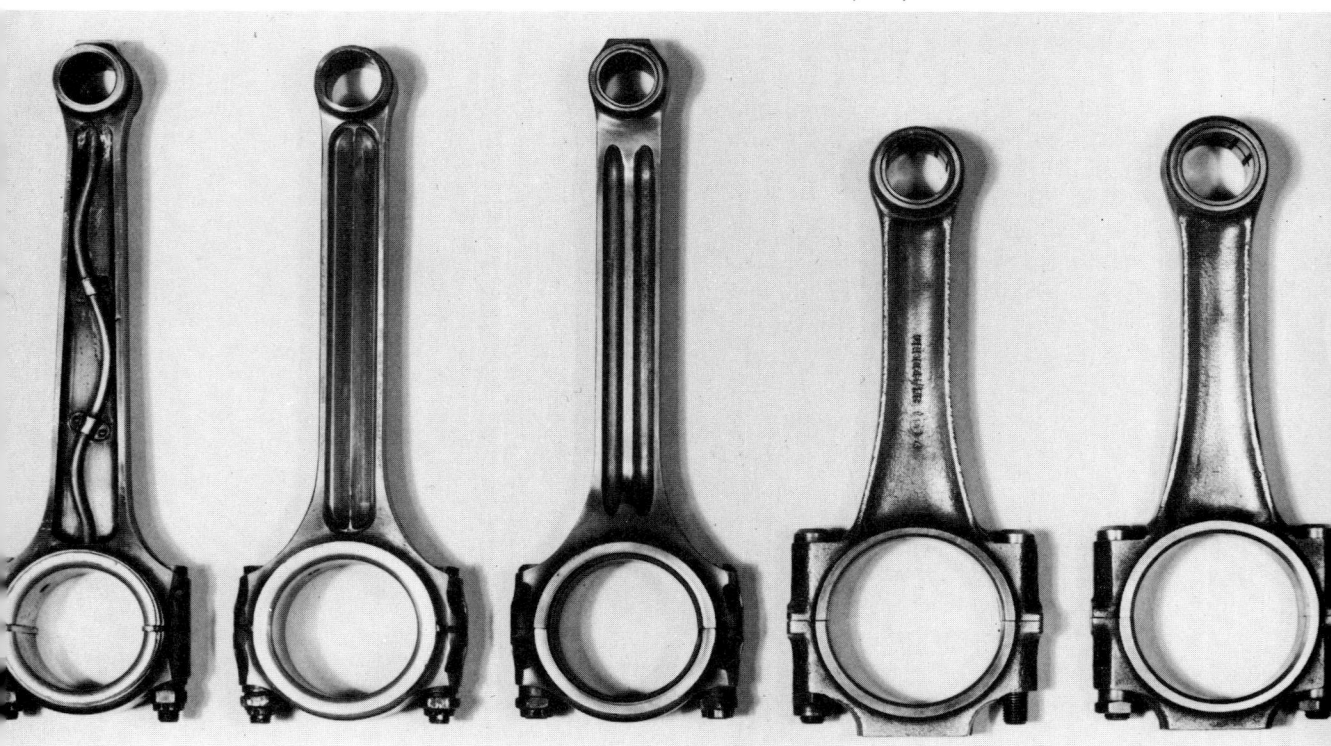

Examples of Rolls-Royce Engineering Style

Silver Ghost connecting rod manufacture, showing 14 machining stages. They are not, however, laid out in the correct sequence!

Lincoln V. 12 K series 1932-1938 connecting rod, an unmachined forging of slender appearance.

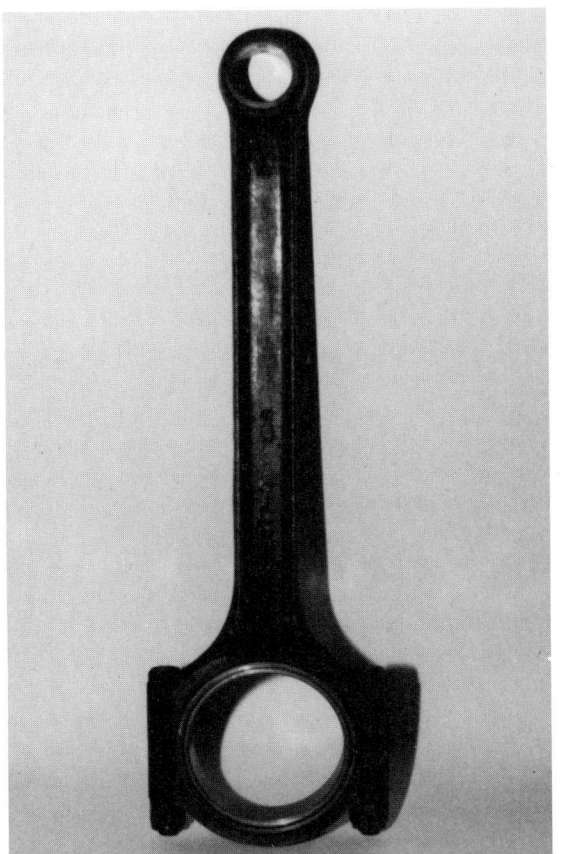

Silver Ghost big end bolts. Early type is on the right, with its coarse thread. A late type is on the left, with a fine thread and hollowed from head to the start of the thread.

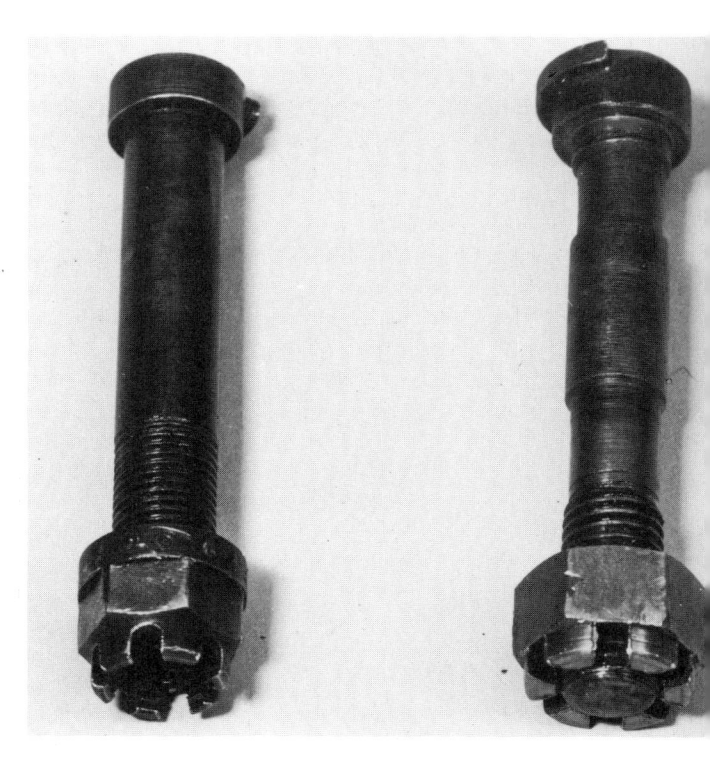

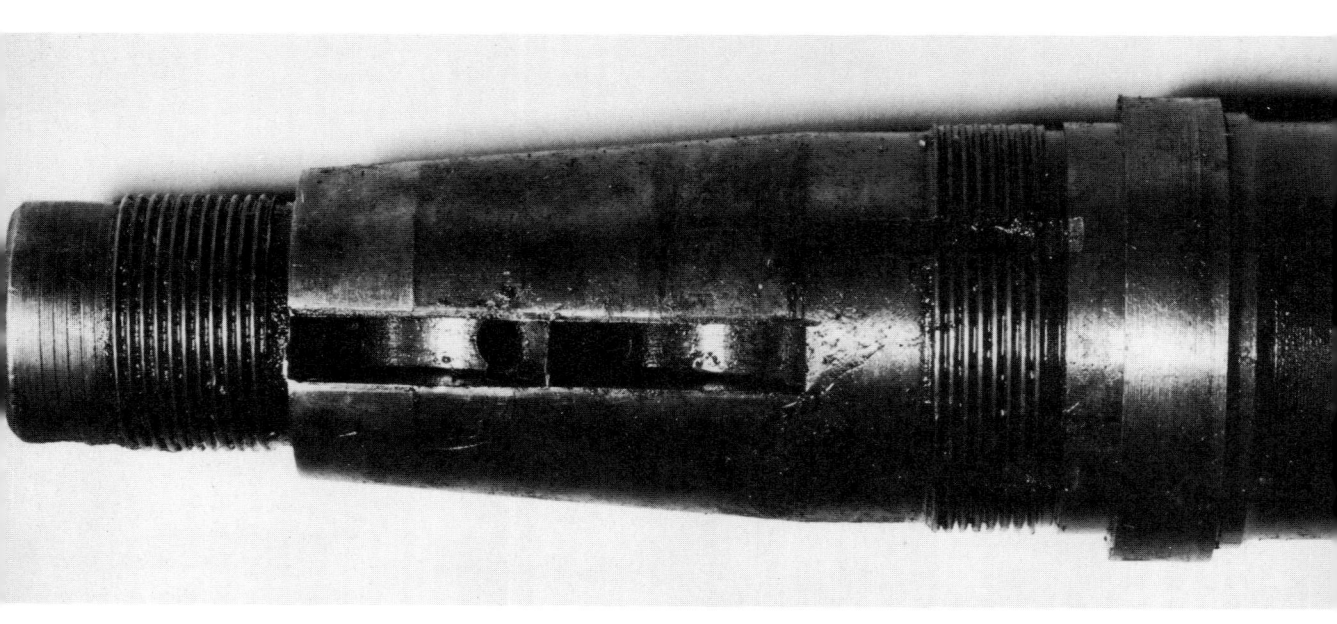

Early Silver Ghost axleshaft embodying a total of four Woodruff keys and two drillings, presumably to assist key removal.

Wraith/Dawn connecting rod. It is similar to the pre-war design in general proportions, but now the central rail is semi-circular, thus avoiding the sharp edges of the previous diamond pattern. The big end bearing is now a standard thin wall shell with no thrust faces. The enlarged 4.9 litre engines, and a few later 4.6 litre units, were given enlarged big end bolts spaced further apart in order to clear the big end bore.

Figure 156 (d) shows the first V8 rod utilised in the Bentley S.II, early Phantom V and Silver Cloud II models. The previous costly, but elegant, all over machining was now totally abandoned, and the rod appears unduly slender adjacent to the head of the big end bolt. The author has yet to see a failure at this point, and the only – rare – fatigue failures encountered have occurred half-way along the length of the rod. The final connecting rod in figure 156, (e), is that of the Bentley S.III and equivalent Silver Cloud and Phantom V/VI models. The small end diameter was considerably enlarged, and this design is also common to the short stroke Silver Shadow engines, built up to late 1970.

Early Silver Ghost Pinion Shaft to Propellor Shaft Flange

It is amazing to find that there is a total of four deep Woodruff keys (two on each side) together with two holes drilled right through the shaft. The purpose of the latter is not clear, unless to assist in removing two of the keys. The shaft is seriously weakened as a result of the mass of material which has been removed. A taper, if accurately made and sufficiently tight, will transmit a large amount of torque, without any positive locking device. However, a single shallow key would ensure reliability without unduly weakening the shaft.

Method of Attaching Axle Shaft to Hub on All Cars Built 1920–1939

The axle shaft has an exceedingly shallow spline, thus causing minimum weakening of the shaft, but a separate driving dog is interposed which in turn locates in splines inside the hub. This is a typical example of Rolls-Royce complication; a further wearing point is interposed, and there is additional cost in manufacture. The driving dogs are troublesome in service and soon give rise to clonks when the inevitable wear takes place.

Later Rolls-Royce axleshaft and hub driving dog. Note the extremely shallow splines.

Piston Gudgeon Pin

Gudgeon pins on the V8 engines are offset in the piston boss by $\frac{1}{16}$ in., the greater distance to the side opposite thrust. This feature was incorporated to assist in the reduction of piston skirt rattle.

Nuts and Bolts

Henry Royce applied his mind in the early years to the question of nuts and bolts. The British Standard Fine (B.S.F.) nut depths, although perfectly satisfactory for all normal engineering work, provided an insufficient margin of safety for Royce. He increased the depths of his nuts to approximately 1.16 times that of a standard nut and also utilised a square head for bolts which would lock against a ledge on the component, and thus avoid the necessity for two spanners when tightening.

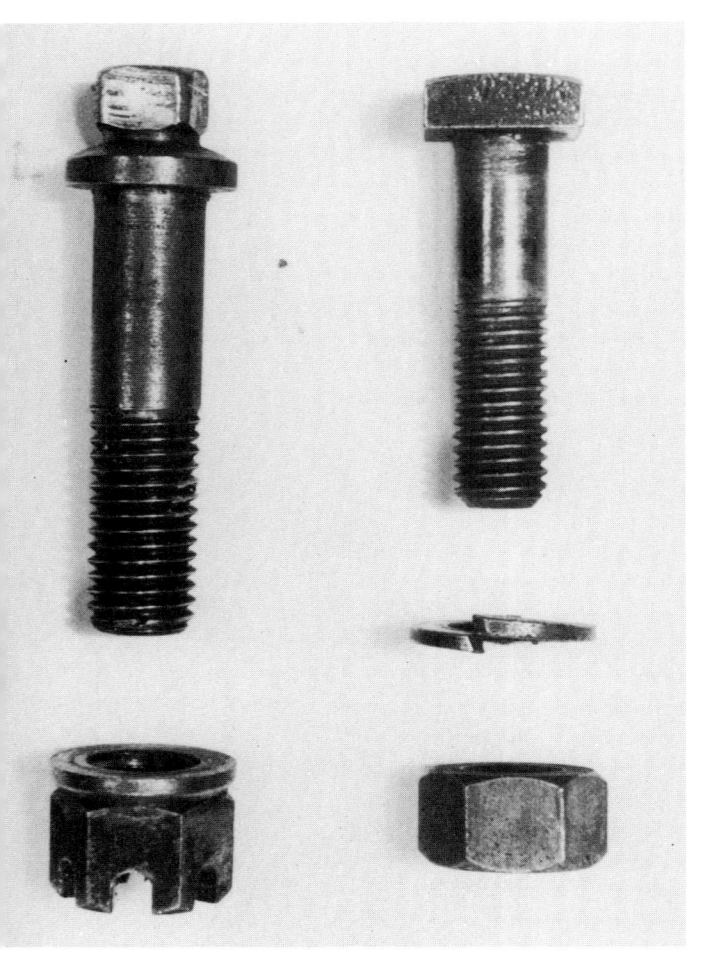

Left: *a Bugatti nut and bolt. Note the integral washer.* Right: *Rolls-Royce nut and bolt, with a square head and separate washer.*

Another designer reviewed in this book – Ettore Bugatti – also took an interest in this subject. His artistic sensibility was no doubt offended by the thought of hexagons gouging the component against which they were tightened. He designed an elegant, if expensive, range of nuts and bolts for his cars which combined an integral washer while the bolts were headed with a small square. Bugatti naturally specified the metric fine thread and retained, like Royce, a deep nut – which in any case was standard metric practice – approximately 1.20 times the depth of an equivalent B.S.F. part.

The Sunbeam Company revealed their European relationship by retaining the metric system until the end in 1935. Lanchester appeared satisfied with the standard British system.

The cost of undue complication and the benefits of using outside component suppliers are succinctly revealed by the following list of components relating to one area of the chassis:

No. of Components of one Front Brake Assembly

The complication and cost resulting from the pre-war design philosophy, and the enormous savings which accrued later from the adoption of proprietary equipment can be readily seen by reference to the schedule of components used in the two designs of braking systems.

	Bentley $3\frac{1}{2}/4\frac{1}{4}$ litre 1934–1939 (R.-R. system)	Bentley R Type 1953–1955 (Girling system)
Split cotters	10	5
Clevis pins	8	6
Bushes	8	–
Rivets	59	40
Locktabs	6	–
Studs	14	3
Bolts	36	8
Nuts	50	12
Washers	32	5
Springs	5	6
Yokes	4	1
Distance pieces	2	6
Rods	3	1
Main shoes	2	2
Rocking shoes	2	–
Liners	4	2
Drum disc	1	1
Drum rim	1	1
Brake cam	1	–
Brake cam arms	2	–
Dust shields	2	6
Carrier plate	1	–
Flange	1	–
Back plate	1	1
Housing	1	1
Levers	2	1
Adjuster	1	1
Peg	1	1
Trunnions	2	–

Examples of Rolls-Royce Engineering Style

	Bentley 3½/4¼ litre 1934–1939 (R.-R. system)	Bentley R Type 1953–1955 (Girling system)			
Key	1	—	Wheel cylinder	—	1
Collar	1	—	Cups	—	2
Plungers	—	2	Pistons	—	2
Screws	—	3	Bleeder screw	—	1
Brackets	—	4	Links	—	2
Gaskets	—	2	Pivot	—	1
Seals	—	3	Plates	—	2
Strips	—	2	Total no. of pieces	264	137

APPENDIX I

Rolls-Royce, Sunbeam and S.S. Cars

Rolls-Royce Ltd

The design philosophy set down by F. H. Royce in the early years was one of refining existing ideas rather than of innovation, and this philosophy his disciples rigidly followed. The combination of this tradition, together with the brilliant commercial direction of Claud Johnson and assiduous cultivation of the Rolls-Royce mystique, was to place the firm in a pre-eminent position within a few years of foundation. The business became very much larger in size than any of its British or European competitors by the time of the early '30s depression, a period which the Derby firm withstood far better than any of its contemporaries. The mystique and prestige carried it through a period of design stagnation during the late '20s and '30s.

A serious review of the entire motor car operation forced the development of a new range of technically up-to-date and simplified designs which would have appeared in 1940 or 1941 had not war intervened. These new cars meant that the company was very well placed during the early post-war period and logical development continued, by which time serious opposition to its traditional market had withered and died.

From 1946 to 1965 the firm, now based at Crewe, could sell all the cars that it could comfortably make, with the exception of one or two bad seasons, the latter largely brought about by the political climate – 1962 for example. An objective assessment is difficult to make by an outsider many years after the event, but it seems reasonable to suggest that Rolls-Royce personnel were obsessed with looking over their shoulders, and worrying about the efforts of others, instead of getting on with some innovative thinking of their own. The firm was, in fact, hoarding large numbers of first class, often brilliant engineers, and who must have been capable of doing much better. One can conclude that there must have been considerable frustration amongst the staff.

It is difficult to understand why some of the Rolls-Royce problems, such as handling and ride, should have proved so intractable, taking into account the weight of brain power, endless discussion and vast expenditure in the experimental department that were brought to bear for so many years. Research among surviving papers reveals a surprising amount of personality problems, a facet which at the least would have an effect upon general efficiency, and at worst could act as a serious brake on the firm's progress.

The decision to operate a design bureau at Royce's home, 150 miles away from the works, would not make good sense, exacerbated still further during the winter when the outfit moved to the South of France, making contact with the works virtually impossible. There is only one place for a design office and that is on top of the job in the works. We know that Royce was continually frustrated because he was rendered powerless to force the pace, while Derby in turn were put to the expense of an additional section in the general drawing office in order to make the Royce office designs workable.

The B.S.A. concern took a similar course in recent times by transferring their design department to Umberslade Park, Hockley Heath, some ten miles away from the works in Small Heath. The result was that the staff could never find each other, and an enormous amount of time was wasted in travel between the two points.

The death of Royce in 1933 caused an immediate closure of the remote offices, and one imagines that a considerable increase in efficiency resulted.

This period saw the start of a move towards the employment of bought out proprietary components. The first such items concerned the fitting of S.U. carburettors to the Bentley, shortly to be followed by the Borg & Beck clutch and Marles steering gear. This gradual process should have made for a considerable reduction in the size of the Rolls-Royce drawing office. The fact was, however, that the new parts did not readily fit into the scheme of things, and the old specialists were jealous of their position, insisting on Rolls-Royce standards, which meant the most severe inspection system and a virtual re-design of the components concerned; this process took several years.

Rolls-Royce, circa 1935, were employing a total of some seven thousand people, including a staff of probably five hundred. The same engineers doubled up on aero and motor car work until 1937, when a split took place following a reorganisation upon the appointment of Hives as General Manager. It is interesting to note that the De Havilland aero engine staff under the leadership of Major Frank Halford and Dr E. S. Moult were constantly amazed by the lavish expenditure which they noted in the Derby experimental department compared with their shoe-string budget at Edgware. Of course this shortage of funds resulted in the occasional taking of risks which would not have been sanctioned at Derby. Nevertheless, success per pound sterling expended by De Havilland must have been several ratios higher than that achieved by Rolls-Royce, the natural benefit enjoyed by keeping the business small.

Sunbeam Motor Car Co. Ltd, Wolverhampton

The Sunbeam concern attained eminence prior to the First World War following the arrival of Louis Coatalen as designer to the Company in 1910. A Breton by birth, he attained fame as a very young man with his ability to obtain higher power outputs than most. He combined a strong personality with qualities of leadership and the ability to pick able assistants. He possessed powerful ambition and quite seriously pursued the idea of marrying into the Chairman's family of each firm that employed him; he was thwarted at Humber and Hillman, but succeeded at Sunbeam!

Considerable growth took place during the First World War with a successful foray into the aero engine and airship engine business. This division

Louis Coatalen.

Sunbeam Dawn experimental car, 1933, outside the factory recreation ground.

was, however, abandoned shortly after 1918, and cars once again became the sole preoccupation. The Sunbeam company became the largest British high-grade maker during the '20s, catering for 14 h.p. through various steps to 30 h.p. cars.

An expensive, but successful, racing and record breaking programme was undertaken, in this field there was considerable liaison with the Parisian firm of Darracq. Sales, and private car design, undoubtedly benefited from the competition successes. Coatalen left Wolverhampton for the Paris design office in 1926, and it is widely held that design stagnated and the impetus withered with his departure. This would seem to be just a little inaccurate, because production and success continued unabated until 1930.

The depression then took a serious toll of sales, while the Directors sanctioned expenditure on various lost causes. The design of high-speed coaches, heavy commercial vehicles and trolley buses was undertaken, only the latter being partially successful. The expense of this work at such a time, together with an abortive land speed record car for Kaye Don, proved disastrous. The last venture resulted in a successful action for damages by Kaye Don, adding further difficulties.

However, in car design the company kept abreast of its competitors in the matter of road performance, handling and appearance, right until the end,

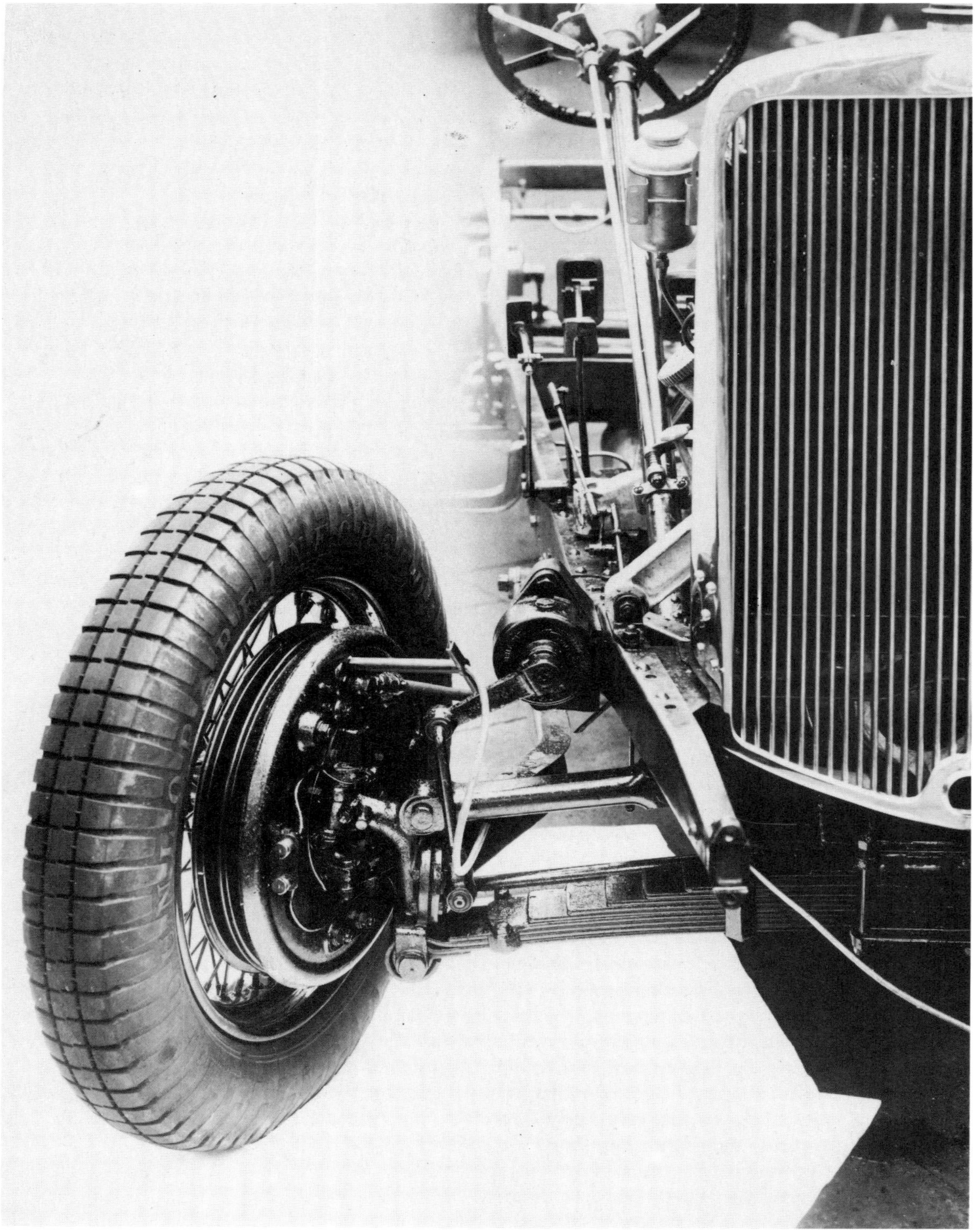

Sunbeam Dawn independent suspension. The upper link is too low to achieve sufficient rigidity. This chassis would appear to be undergoing a brake-bleeding operation. Behind is the pre-selector gearbox linkage.

and was amongst the first to adopt hydraulic brakes, while a new 12 h.p. car aimed at a wider market in 1934 bristled with advanced features, including independent front suspension, a pre-selector gearbox option, and an aluminium cylinder block with wet liners – the latter having been seen before on a four cylinder of 1922. However, the new car – the Sunbeam Dawn – possessed shortcomings due to lack of development which were never rectified due to the imminent backruptcy.

The company's commercial direction would certainly seem to have been misguided, one manifestation being the large staff that was retained busily engaged on the drawing boards until near the end. In 1934 the design staff, under H.C.M. Stevens, comprised four senior men who classed themselves as designers, together with eight draughtsmen, four semi-skilled youths, three girl tracers and two apprentices, the foregoing to be found in the main drawing office. Additionally, separate offices housed two to three jig and tool men and two to three body design specialists. A very small drawing office in the pattern shop was also maintained. Clearly, the overhead costs must have been unbearable at this period, while the staff were apparently unaware of the company's chaotic financial situation.

S.S. Cars Ltd

The products of this concern do not, at first sight, appear to warrant comparison with those of Rolls-Royce. The rapid development of the S.S. marque from humble beginnings in 1932 was so spectacular, however, that their larger offerings were actually affecting Bentley sales by 1938.

The history of S.S. Cars, later Jaguar, has been told many times, but some aspects of the philosophy that lay behind the business appear to have been overlooked. The meteoric success was, of course, entirely due to the dynamic single-minded drive, commercial sense, and highly developed artistic flare of one of the founders – the late Sir William Lyons.

Initially, several seasons were spent on the production of distinctive coach-work for popular small chassis. Lyons then prevailed upon Captain (late Sir) John Black, the recently appointed Managing Director of the Standard Motor Co. Ltd, to produce special lowered chassis for their 9 h.p. and 16 h.p. models exclusively for Lyons, the car to bear the S.S. monogram.

The cars were ready for the 1932 season. Despite the acute depression, and the fact that they were woefully depressing in performance, they somehow struck a chord. The first year's sanction of some five hundred cars were cleared without much difficulty. These first cars were slightly weird in appearance, apparently due to the fact that Lyons was struck down by illness at a crucial moment in design. The following season saw a great improvement in appearance, and the car was gradually developed while production built up. New models appeared, but Lyons could see that this type of hybrid car suffered from limitations, due to the fact that it was based on a cheap mass production model.

The design office at Standard, busy with its own new models, could not cope with the design of an entirely new car, that Lyons had in mind. He was eventually pressed into starting his own technical department, and Edward Grinham, Standard's chief engineer, suggested that W. M. Heynes, a designer with the Humber-Hillman combine, should be approached. Heynes joined S.S. in April 1935; he was given a very small office, and engaged one draughtsman.

The first $2\frac{1}{2}$ litre S.S. Jaguar was designed from scratch, and the first cars completed in comfortable time for their appearance at Olympia in October 1935. Serious production commenced immediately. A new $1\frac{1}{2}$ litre car was also prepared at the same time, which admittedly required less work since it was to continue with the side valve Standard 12 engine for a further two seasons. The secret of success was due to the choice of a first-rate technical man, very heavy reliance on outside suppliers, continual help from the Standard concern and a very tight budget.

The range was augmented with a $3\frac{1}{2}$ litre model for the 1938 season, selling for £445. This car was as fast as the $4\frac{1}{4}$ litre Bentley, quiet, handled well and was better looking. The body was every bit as good

S.S. Jaguar 3½ litre, 1938–1939. The first all-steel body, jigged and welded together at the S.S. works. The photograph was taken outside the old offices in Foleshill.

S.S. Jaguar 2½ litre saloon. Designed in 1935 and probably the first car built.

as the Standard Steel Park Ward Bentley, and had more room within. Of course, this car suffered from fragility of some components, and owners had to put up with floppy fittings which would not have been tolerated at Derby. Nevertheless, the fact that you could buy three S.S. Jaguars for the price of one Bentley was a cause of some amazement. The 2½ litre car, priced at £395, represented even more remarkable value, since it was virtually identical to the 3½ litre, and the works cost of both cars would be almost identical.

The design facilities at Coventry were gradually built up and some five or six draughtsmen were employed by the time war broke out. Heynes was then running three experimental cars, and was given sanction to purchase a Frazer-Nash B.M.W. for the purpose of technical examination, the first such acquisition. It is interesting to note that a Rolls-Royce Silver Wraith was purchased by the company in about 1950, and a Bentley Mark VI sometime later. The Jaguar engineers were not particularly impressed, finding the ride inferior to their own new cars embodying torsion bar independent front suspension. The Rolls-Royce brakes also came in for criticism, Walter Hassan, Heynes's deputy, totally exhausting them before reaching Worcester on the usual Jaguar test route. This factor indicates, perhaps, the different type of driver that bought Jaguar cars. The Rolls-Royce braking system feels so good in normal usage.

Appendix II

Notes on some of the engineering personalities

A short description of the personalities, work and position of various engineers concerned with the companies reviewed in this work will be of interest. A proportion of them, often the most valuable, have remained obscure, with but little acknowledgement for excellent work performed in the development of the various cars described.

Sir Henry Royce, Bt, 1863–1933

Despite the lack of a good general or professional education, Royce educated himself to a high standard and had a complete and incisive grasp of all engineering matters. He was strongly motivated in his early days and, with a wealth of experience in his later years, remained of great value to Rolls-Royce Ltd. He was sometimes a little indecisive with regard to ideas on future policy and could vacillate over major design decisions.

It is commonly stated that serious ill health in 1910 caused his banishment to warm climes and his divorce from the day-to-day stress of factory life. He nevertheless lived a further 23 years in what must have been idyllic conditions, reaching almost seventy years of age. His later life consisted of unruffled days spent in the South of France in winter and Sussex in summer, rising late, taking exercise, enjoying stimulating design discussions with like-minded colleagues and interesting friends in the evenings. He was cossetted by a doting nurse, financially very comfortable and the recipient of many honours, culminating in a baronetcy. What more could a man ask?

Interesting light is thrown on his character by this memo from Royce to Walter Hives, then his Experimental Shop Manager at Derby in 1929. The memo is rather bad-tempered, indicative perhaps of the frustration caused partly by the self-imposed impracticality of trying to run a design office 150 miles away from the works. (It is surprising that such an observant and meticulous man did not take the trouble to spell Alfa Romeo correctly.)

Hs. from R. R3/M8.11.29
c. Sg. Wor
c. Rg. By. X.3054
X.4482

RE – <u>ALPHA ROMEO</u>

I do not consider we ought to spend £900 in purchasing the 2 litre car, of any make at the moment.

Why? Because we have so much to do that we know is good and are not stopped for ideas.

You will remember that we cannot get along quickly enough with the things already instructed from WW, so *we must settle down to some real work* until we begin to get slowed up for want of ideas, or difficulties which we do not understand.

It took me 2 years to get vertical shutters on the 20 HP. so as not to look like an Essex, but when we had made the improvements the Essex was already there.

It has taken several years to get 3″ extra length of frame for the 20 HP. to avoid short stumpy scuttles and high seats and steering – still not available.

We have brake problems that require experimental explanation which has taken too long.

We are suffering from too little maximum HP. for the 20 HP, parts for which you have had for over a year, (now moving by pressure from WW. and London).

A great deal of work is done at WW, which produces no results. Either we are stupid, or you have not the capacity I imagined.

It should be realised that parts should not be instructed by the Exp. Testing Dept. unless referred to those responsible for the design – i.e. reports on the design should be sent to the designer with suggestions for curing any faults found – because we cannot afford the present way for time and money, and it crowds out other work.

I want to get the work sent from here attended to promptly, or save the Company the WW. expense. Probably much improvement will result from specialising, and dividing the Exp. Dept.

Lord Hives, C.H., M.B.E., 1886–1965

Ernest Walter Hives was born in 1886, in London. He joined C. S. Rolls & Co. as a tester in 1904, left to join Napier and finally joined Rolls-Royce in 1910 as Head Tester, carrying out record-breaking work with the Silver Ghost at Brooklands and elsewhere. He became Experimental Shop Manager after the 1914–1918 war and remained in this post until 1937.

He soon became a powerful figure and exerted a great influence over policy while in this post. He was, in particular, responsible for the specification of the first $3\frac{1}{2}$ litre Bentley. His firmly-held views on this project were, from the outset, utterly sensible and logical. It seems that Sidgreaves, the Managing Director, was grateful for Hives and always seems to have grasped the drift of Hives's reasoning.

There is no doubt that such a man was bound to reach the top. He applied for the post of General Manager upon the death of Wormald in 1937 after seeking the views of his closest colleagues. Following appointment to this post, and no doubt after consultation with the Air Ministry, Hives's first decision was to split the aero engine and motor car business into separate functions. Rapid advancement followed, culminating in elevation to the peerage and the managing directorship of Rolls-Royce.

Hives did not hold professional qualifications; he was, nevertheless, an outstanding practical engineer with a complete understanding of all the facets involved in this fascinating subject. Hives was always highly suspicious of designers, who he regarded as a confounded nuisance, instead subscribing to the view that the evolution of cars and engines should gradually advance by way of development and test, which would take place in the experimental department. He did, in fact, always enjoy the facility of a good drawing office in this department during his term as Experimental Chief. He died in April 1965.

Arthur John Rowledge, M.B.E., 1876–1957

He was born at Peterborough, a few miles away from the birthplace of Royce and 13 years later, and was the youngest of a family of four. His father was a proprietor of a small building firm and his mother came from farming stock. Arthur Rowledge followed the artistic flare which his father possessed and was intensely interested in mechanical things from an early age, obtaining a Queen's prize in science subjects at school. He took an apprenticeship with Baker and Perkins of Peterborough, but does not appear to have received professional training. He gained further experience with printing and general engineering firms, including time as a draughtsman.

He then joined Napier in 1901 as a designer and left in 1905 to take a post with Wolseley, eventually becoming Chief Designer. He rejoined Napier in November 1913 as Chief Draughtsman but soon advanced to the position of Chief Designer. He was responsible for the world famous Napier Lion aero engine, work on this project commencing in 1916. The success of this unit was such that it remained in production with constant development for some twenty-five years.

Rowledge then evolved the post-war 40/50 Napier car, an advanced design embodying aero engine practice. This car was not particularly successful but development work seems to have been suspended. Rowledge left to join the design staff of Rolls-Royce as personal assistant to Henry Royce, with particular responsibility for aero engine design. He made an outstanding contribution to the success of Rolls-Royce and was directly responsible for the famous Kestrel, R type and

Appendix II

Lord Hives (left), with Dr Llewelyn-Smith on the right.

Merlin in addition to exploring single sleeve valve engines. He possessed the valuable ability to make good designs which could be built with existing plant in the works and was a good improvisor.

He was regarded by his colleagues as a genius, but was quiet, unassuming and shunned publicity. He had no great ambitions, was perfectly content to get on with his aero engine work, and found his greatest pleasure in relaxing with his family. He was called in to assist with motor car works from time to time and is remembered for refining and adopting the Hispano brake servo for use on Rolls-Royce four wheel brakes in 1924. He retired at the age of 69 in January 1945 having been chief of the design staff at Derby for some twenty years.

R. W. Harvey-Bailey, 1878–1962
(by Alec Harvey-Bailey)

Educated at Tiffins in Kingston-on-Thames, he gained an exhibition to King's College, London, where he studied civil engineering. In 1895 he was a Jelf Medalist and in 1896 became an A.K.C. (Associate of King's College). After fourteen years in the infant automotive industry, where he was a pioneer in front driving, particularly heavy steam traction, he joined Rolls-Royce as a designer with section chief status in the spring of 1910.

He worked with Royce at Derby on the torque tube axle and spent long periods at Le Canadel, Royce's summer home in Provence, during the firm's formative years. His work for Royce included the four-speed gearbox for the Austrian Alpine Trials cars and, following this, the big brake system.

During the 1914–1918 war he concentrated on aero work and at Royce's direction designed the Falcon engine. He also took a wide interest in the many technical problems thrown up by aero engines. In a post-war reorganisation he was appointed Chief Technical Production Engineer. As well as running the Detail Drawing Office he acted as an interpreter of the requirements of Royce's design office at West Wittering. His work took him into the field of non-conformance problems, special customer requirements and service issues on both chassis and aero engines. There were times when he took on design tasks which included the Peregrine car. Although not officially a designer, he retained his love of design and for many years had a drawing board in his office on which he produced schemes to handle difficulties on production and in service.

Shortly after Hives became General Works Manager he carried out a re-organisation, splitting aero and chassis into two divisions. In 1937 By, as he was known, was appointed Chief Engineer, Chassis Division, charged with rationalising chassis activity and producing a new range of cars. There followed two exciting years when the Bentley V was brought to production status and a new range of Rolls-Royce and Bentley motor cars was under development. This included the straight-8 engine to power a Phantom III replacement and a new high-performance Bentley. He was an enthusiast for the concept of the streamlined performance car and employed Georges Paulin on the Corniche. Had the war not come the Bentley V and Corniche would have been announced at the 1939 Motor Show and the first truly rationalised cars, including the straight 8s, would have been at the 1940 Show.

However, on the outbreak of war Hives recalled By, to aero work, where he once again became a pivotal figure in the expansion programme covering quality, repair engineering and service failure investigation. A major task was the use of non-conforming material without affecting product integrity and schemes to simplify production parts.

He was not then a young man but worked with enormous energy not only at Derby but also with the other major factories producing Merlin engines. He retired on 31 December 1945, less than two months short of his sixty-ninth birthday.

Donald Bastow, B.Sc. (Eng.), C.Eng., M.I.Mech.E., b. 1909

Donald Bastow was born on 8 June 1909, the son of an accountant. He was interested in mechanical things from childhood. He obtained a B.Sc. with first class honours at London University, joined the Daimler Company in 1929 as a post graduate pupil and stayed for three years, gaining experience in all departments.

He took a position as draughtsman in Sir Henry Royce's design office at West Wittering and is, in fact, the last surviving person to have worked with

Royce. Transferred to Derby on Royce's death, he was given the task of laying out the first i.f.s. system (Phantom III). He then specialised in the field of chassis and suspension design, becoming nationally recognised, and was transferred to tank design during the war.

He joined Lagonda in January 1944, moving south for domestic reasons, and became personal assistant to W. O. Bentley. He advised on the design and assisted in the development of the post-war 2.6 litre model, but did not design the suspension system on this car. Bastow confirms that the shortcomings would have been made good, but in the event Lagonda passed into the hands of David Brown before the car was fit for production; nevertheless, David Brown considered the car satisfactory and built it for some nine years without modification.

Bastow stayed as personal assistant to W. O. Bentley until 1950; in this period he assisted with work on a still-born Armstrong-Siddeley, which was designed on a consultancy basis by the Bentley team. Bastow then spent two years at B.S.A., starting a specialised research division for this concern, leaving in 1952 to become Chief Engineer of Jowett Cars Ltd. He stayed until the end of car production. Has since held various senior engineering posts and for the past seven years has operated as an independent consultant. From 1959 to 1960 he was President of the Automobile Division of the Institute of Mechanical Engineers.

Geoffrey Bastow, b. 1907

Born on 13 May 1907, he was similar in many respects to his younger brother Donald. He served an apprenticeship with Vauxhall, with whom he stayed until he was transferred to the Rolls-Royce tank division in 1942; he thus worked with his brother for a period. He joined Rolls-Royce motor car design staff in 1945, staying until his retirement in 1972. He carried out a great deal of good work for Crewe, often being asked to solve irksome problems ranging from suspension work to heater controls.

Quiet and unassuming, he does not appear to have received sufficient recognition by the authorities at Rolls-Royce.

H. Ivan F. Evernden, d. 1980

He gained a B.Sc at King's College, London and joined Rolls-Royce at Derby in 1916 as a draughtsman on the jig and tool section. He then worked with Royce at West Wittering in 1921. He possessed a highly developed flair as an artist and became responsible for coachwork design and liaison between the factory and the coachbuilders. He produced some excellent drawings of interior layouts and so forth. He remained at Rolls-Royce until retirement, where he is particularly remembered for the R Type Continental produced to his designs. He was not considered to be a great engineer, sometimes choosing a complicated path to solve a simple problem. He was also apt to appropriate the ideas of others and was thus not particularly popular with some of his colleagues.

He made statements in an article about his chief, Henry Royce, which do not appear accurate. He claimed that Royce was no draughtsman, which cannot be true. It is almost certain that Royce would have undertaken a great deal of drawing board work in his early days with the design of cranes and the early cars; by the time that Evernden knew him he would have passed this stage. We do know that Royce possessed a fine set of drawing instruments, which are still in existence. Evernden mentions that keys were never used, but Woodruff keys were definitely used on the early cars, cutting deep into shafts, as described earlier. Failures would have been experienced and they were abandoned in the 1910 re-design. Evernden also states that full nuts were never used. This is not so.

B. I. Day

He was Chief Engineer at Sheffield – Simplex, responsible for designing their luxury car which appears as if it may have severely threatened Rolls-Royce. Mercifully, Earl Fitzwilliam, the patron of the firm, withdrew support and production ceased in 1921.

Day joined Rolls-Royce in 1913; it is thought that he was head hunted, probably by Claud Johnson. He formed a design office at Derby and carried out a great deal of aero engine work, joining the Royce design office at West Wittering in 1921.

He then specialised in chassis work and is remembered as being an outstanding engineer and designer.

A. G. Elliott, C.B.E., F.R.Ae.S.

He served for a time with Napier before leaving to join Rolls-Royce in 1912, and became head of Royce's personal design team in 1917, but concentrated on engine work. He is remembered as a very fine draughtsman but not for being particularly innovative. He was extremely ambitious; he married Royce's secretary and thought out his career moves as if playing a game of chess.

He left West Wittering in December 1931 during one of Royce's bouts of illness. He imagined that the Chief might not recover and so took off for Derby, where he seems to have appointed himself as Chief Engineer. There was some kind of vacuum at Derby; possibly Managing Director Sidgreaves was partly out of control of the situation, being of course based in London and not an engineer himself.

There is some extremely vicious correspondence in existence between Elliott and his colleagues, which belies the team spirit of which we have heard so much.

Elliott transferred to the aero side before the war and ultimately became Managing Director in 1951, Vice Chairman in 1955 and was Vice President of the R.Ae.S., 1963–1965.

Stuart Tresilian

He joined Rolls-Royce as an apprentice in the early '30s and concentrated on engine design. A theory man, he produced a design for an all-aluminium engine, embodying three valves per cylinder, thus following the style of his Type 35 Bugatti, which he ran until exchanged for a Type 55, the ultimate sports car of the period. The Type 35 was dismantled by Rolls-Royce who measured, weighed and carried out a full investigation of the power unit. The aluminium engine designed by Tresilian embodied an open cylinder block top face with wet liners, presumably spigoted into position. He then did a considerable amount of work on the Phantom III engine.

He joined Lagonda in about 1936 and persuaded W. O. Bentley that he must go in for short strokes, despite the British taxation system. The V12 Lagonda engine was the result, differing from the Phantom III in embodying overhead camshafts; it nevertheless bore certain similarities to the Rolls-Royce power plant. Although much smaller in capacity, the power output was almost as high. The Lagonda engine would have afforded very serious opposition for Rolls-Royce if sufficient funds had been forthcoming to develop this imposing car. Alan Good, the Chairman of Lagonda, obtained financial backing of £250,000 from a private individual in order to assist with production. However, the money was lost, and Tresilian left Lagonda to become Chief Engineer of Bristol Cars.

W. A. Robotham, 1899–1980

A particularly prominent personality at Rolls-Royce during the '30s and '40s.

Born in 1899 and educated at Repton, he joined the Army, obtaining a commission just as war ended in 1918. His family were solicitors but he joined Rolls-Royce as a Premium Apprentice, initially in the experimental department under Hives, becoming chief chassis experimental engineer in 1930 and senior car designer as deputy to Harvey-Bailey in 1937.

He was a forceful character and a good leader who was quick to grasp engineering problems, although not a designer. We learn from Donald Bastow that he was incapable of reading a drawing and had to have 'the part in his hand' to comprehend matters properly. He was capable of arrogance and occasionally made unjustified remarks about the work of other manufacturers, some of which he must have lived to regret.

He fully supported the rationalisation of models planned for 1940 and went on to propose drastic changes of policy for the post-war years, including serious consideration of attacking the volume 8 h.p. car market, the medium size 12–14 h.p. range and also the 2–5 ton truck business, all in addition to Rolls-Royce's traditional expensive car market.

These counsels did not prevail and Robotham was seconded to the Ministry of Supply as Chief Engineer of the tank design unit. He returned to Rolls-Royce as Chief Engineer of the car division

W. A. Robotham

and was appointed a Director in 1949. His final appointment involved a transfer to Shrewsbury as Managing Director of the oil engine division, based in the old Sentinel steam wagon factory.

Dr F. W. Lanchester, 1868–1946

Undoubtedly this country's greatest engineering scientist in the pioneering years, he was intellectually brilliant, and pursued attention to detail in the manner of Royce. He was responsible for much theoretical work in the field of aeronautics and pioneered many radio developments, and is particularly remembered for improvements in road vehicle suspension developments. Although more successful as a family man than Royce, he found difficulty in establishing a good working relationship with all but a few of his contemporaries, a problem greatly aggravated with old age. A very poor financial manager, he spent his last fifteen years in conditions of quite serious poverty. Leaders of the motor industry clubbed together in order to buy him a Standard 8 car in 1939, since he was without any form of transport or the wherewithal to provide it. He was a founder of the I.A.E. and recipient of many engineering and scientific prizes, but did not receive any official honours.

He was retained by the Daimler Co. for many years as a consultant, but his opinions were seldom sought after 1925. A promising relationship began with the Wolseley Co. in 1924. This concern, still a division of Vickers, was plunging into desperate straits after enjoying several years as Britain's largest manufacturer. They lost ground heavily in the early '20s, finding themselves saddled with a wide range of dull cars. It was felt that the brilliance of 'Dr Fred' might provide the answer, and they even contemplated using the Lanchester name in conjunction with Wolseley. This consultancy work, worth £1,000 per annum, was apparently discontinued after one season.

Dr Lanchester, so outstanding in his earlier years, became disillusioned and embittered as time went by, gradually losing his consultancy work due to an inability to strike up a working relationship with his colleagues. He also pursued theoretical solutions to the exclusion of practicality, and developed a total refusal to accept reasonable criticism, a state of affairs which, sadly, tarnished a brilliant career.

George Lanchester, 1874–1970

A gifted, practical engineer and very good draughtsman, he was totally devoted to his elder brother, Frederick, stable and energetic all through a long life. He was President of the I.A.E. in 1943–1944, not perhaps a conceptual thinker in the manner of his brother, but highly respected and exceedingly well-liked by colleagues. He lived to the age of 95 and retained his enthusiasm for the engineering scene until the end.

Jean Bugatti, 1909–1939

Possibly the most gifted of a highly artistic and talented family. Of Italian origin and eldest son of Ettore, he was born in 1909. He exerted influence over design and policy at the Bugatti works from the age of 21, when he persuaded his father to buy two American Miller racing cars with a view to adopting the twin overhead camshaft engine layout; this was in fact virtually copied for the type 50 and type 51 Bugatti engines. Jean Bugatti was responsible for coachwork design and took on the overall responsibility for running the works at the age of 27.

He had no formal training but was well grounded in his trade by having lived at the factory since childhood and thus constantly observed all processes of design and manufacture. He was tragically killed in 1939 while testing a racing car a fortnight before the outbreak of war, aged 30.

Jean Bugatti, aged 20, with a Type 44 in the winter of 1929/1930.

Bugatti Designers

(by Dr H. G. Conway, C.B.E., F.R.Ae.S., F.I.Mech.E.)

Felix Kortz, d. 1927

The first designer or draughtsman employed directly by Ettore Bugatti that we know of was Felix Kortz, who worked with him in his private offic in Cologne from 1907–1909, moving with him when he opened his own factory at Molsheim in 1910. As far as one can tell he remained at Molsheim during the German occupation between 1914 and 1919, while Bugatti himself, being Italian, left for Italy and later France. Kortz rejoined Bugatti when the factory was reopened, and many of the drawings of principal parts of the new car and engine designs bear his signature. In today's terminology, Kortz would have been called the chief draughtsman, but no one at Molsheim had titles except the *Patron*. Examined today, one is struck by the relative complexity of Kortz's designs which, if stemming from Ettore's scheme layouts, resulted in detail drawings of cylinder blocks or complex engine castings covered with miniscule dimensions delicately cored water passages (with wall of 5 mm. or even 4 mm. thickness), threads taken up to shoulders, and all those things designers are told not to do to give the production shop a chance. But at Molsheim the shops and certainly the foundry seemed pleased to achieve the impossible if that was what was wanted. Kortz unhappily met his death in a Bugatti, it is said, on test in 1927. We do not know his age at that time but suppose he must have been about forty.

Edward Bertrand, 1894–1974
Of German nationality, born in Bavaria in 1894 but of French Protestant family origin. He joined Bugatti in 1924 and we see drawings dealing with aspects of the Type 35 Grand Prix car being produced at that time. He started as a junior, as his earliest drawings were for odd tools for the car. His talent quickly developed and he became a very capable engine designer, taking over the leadership when Kortz died. He was one of those rare draughtsmen who could work quickly with a soft pencil to outline some proposal he wanted to make, the results being technically clear and of some aesthetic quality. By 1939 he was in charge of engine work and design of the many special machine tools that the factory made for their work. He went with the Bugatti design office when it moved to Bordeaux in September 1939, but returned to Molsheim after the Franco-German armistice and worked with the occupying Germans at the factory for a while, being deported under accusation of sabotage but surviving to rejoin Bugatti when he got his factory back in 1946. It was an unhappy period then, with Ettore dying and car production being impossible, and he left to join a printing machinery firm in Strasbourg until eventually retiring. He undertook occasional work for Bugatti during retirement and his last job concerned the design of the final Bugatti prototype, a $1\frac{1}{2}$ litre high-performance engine designated Type 252 built during 1963.

Antonio Pichetto
An Italian by birth, he was a designer working for Cappa, who had been Technical Director of Fiat, and ran a consulting design office in Turin, working on, among others, Itala designs.

For reasons that are unclear, Cappa designed with Pichetto a four-wheel-drive racing car, which Bugatti decided to take up. This resulted in Pichetto being transferred to Molsheim and the car becoming the Type 53 Bugatti, using their latest engine, the 4.9 litre 8-cylinder supercharged version with twin overhead camshafts and two valves per cylinder, in fact based on the layout of the American Miller racing engine. Pichetto arrived in November 1931 and worked on the T53 with Bertrand. Later he handled the front suspension for the new touring car, the Type 57, giving it a twin transverse spring independent layout along the lines of that already designed for the four-wheel-drive car. Although two chassis were built with this suspension, Ettore would not let the layout proceed, insisting that a Bugatti should have his design of solid front axle. For the next few years until the war, Pichetto was a principal contributor to the production T57 chassis, the racing T59 and the new Type 64, which was intended to be the replacement for the T57 but was not finished by the time war came. His main work was on chassis design and he was responsible for the 57G 'tank' cars which were so successful at Le Mans and elsewhere. After the armistice in 1940 he joined the small team under Ettore at the Rue Boissière in Paris, designing notably the proposed T73C racing car which Ettore hoped to produce after the war. After the death of Ettore in 1947 the office was closed down and Pichetto joined Gordini, along with some of the workmen engaged on the prototype 73s and other projects.

Noel Domboy, b. 1899
Noel Domboy was born in 1899 and joined Bugatti at Molsheim in May 1932 to work on the design of the new Type 57 chassis. It was not long, however, until he was transferred to work directly for Ettore on the design of the new Bugatti railcar, which was to prove the commercial salvation of Molsheim by

bringing much needed work to the Bugatti factory. Although spending most of his time on the railcar, he still became involved on car work, especially the new 4.5 litre engine. He too went to Bordeaux with the office in 1939, then returning to Paris in 1940, producing, as he has said, many a drawing developing Ettore's sketches of everything from cars to boats and boilers. When Bugatti died he went back to Molsheim, helping on the design of the 251 racing car under the direction of Colombo, who had been engaged to produce it, and being involved in the factory's final fling, a V12 cylinder engine for a car to be known as the Type 451 and intended to compete directly with Ferrari. All work on this project was abandoned when the Bugatti works was taken over by Hispano in 1963.

Hans Ledwinka, 1878–1967

A designer to achieve eminence in central Europe, particularly for his chassis and suspension innovation, was Hans Ledwinka. Born 14 February 1878 in Klosternenburg near Vienna, he received a technical education in addition to a practical tool-making apprenticeship in the Austro-Hungarian Capital. He took a post with the Nesseldorfer Company, a railway rolling stock manufacturer in Bohemia who, in 1897, were experimenting with a motor car built along the lines of the Benz. Ledwinka assisted with this venture, which proved to be the first car built in present day Czechoslovakia. He became chief designer of the automobile division of this concern in 1900, leaving two years later to assist in the design of a steam car with a professor from the Vienna Technical Institute, followed by a short stay in Paris. He was then summoned to return to his old firm in Nesseldorf, where he was given entire control of the automobile division.

He designed a new engine embodying 90° overhead valves contained in a hemispherical combustion chamber and operated by a single overhead camshaft and rockers. It is probable that the first engine was built around 1908, while serious production commenced in 1910. He left to join Steyr as chief engineer during the First World War, where he designed gun tractors and various military vehicles, returning for the third time to the old Nesseldorfer firm in 1921, taking charge again of the automobile plant and now enjoying the title of technical director. The recently formed nation of Czechoslovakia was anxious to further its new found nationalism and distance itself from the German language. Thus the town of Nesseldorf was re-named Koprivnice, while the cars were titled Tatra.

Ledwinka designed a new car in 1921 which revealed astonishing foresight and embodied a tubular backbone chassis together with all round independent suspension, thus anticipating similar layouts evolved by his friend Ferdinand Porsche for Austro-Daimler by six or seven years. This advanced thinking was forced on him by the state of the roads in central Europe of those days. The versatile Ledwinka produced a profusion of designs ranging from small two cylinder cars of 1000 c.c. up to an elephantine twelve cylinder car for the personal use of President Bénès. A range of unorthodox trucks followed, embodying air cooling and four-wheel-drive.

His most famous achievement was probably the Type 77 Tatra of 1934 which combined a monocoque body and chassis with a rear-mounted V8 air-cooled engine. This car embodied an aerodynamic streamlined form of somewhat bizarre appearance with a dorsal fin and almost completely blind rearward vision. The car suffered from faulty weight distribution which affected directional stability, but nonetheless it created a sensation. Direct descendants are built to this day, a continous development spanning more than fifty years and surviving all the political upheavals and national tragedies.

Hans Ledwinka, in appearance not unlike a farmer from the English shires, fared less well. He was directed by the German occupying powers to concentrate on military vehicle designs, only to be imprisoned after the war for a period of six years. He was exiled upon his release, settling in Munich. Those in power apparently had second thoughts and he was invited to return to his old post. He refused, however, and remained in Germany, where he died in 1967 at the age of 89.

William Munger Heynes, C.B.E., F.I.Mech.E., b. 1903

William Munger Heynes.

Born on 31 December 1903 in Leamington Spa, and educated at Warwick School. He originally intended to be a doctor, but joined Humber as a pupil in 1922 and immediately became fascinated with engineering design, discovering he possessed a flair for this work. He remained at Humber as a draughtsman under chief designer Jack Wishart. One of the first projects undertaken solely by Heynes was to convert the vertical overhead inlet valve system used by this firm to an inclined type

Dr Noel Tait receiving a long service award from Bill Heynes.

utilised on the first 16/50 and Snipe models, introduced in 1929. The new engines benefited from the improved combustion chamber shape, and greater turbulence resulted. He was placed in charge of the Humber technical department in 1930, but continually frustrated because his schemes for improving the dull and stodgy cars produced by Humber-Hillman were constantly rejected by an unimaginative management. He produced an independent suspension system for the Hillman Minx in 1935, also rejected.

Heynes left in April 1935 to start a technical department for S.S. Cars, initially to design the first Jaguar models. He remained for the rest of his career, although pressed hard at one time by Sir John Black to join the Standard-Triumph combine as technical chief. He joined the board of Jaguar in 1946 and was directly responsible for the XK engine and the world beating sports racing cars, in addition to all the production saloons, culminating in the XJ6 and V12. He was president of the auto division of the I.Mech.E. 1960–1961.

He is one of our greatest automobile engineers, an ideal team leader and equally at home with chassis or engine work. He retired in 1968 and still farms in Warwickshire.

Dr Noel Tait, Ph.D., M.Eng., b. 1904

Born 31 August 1904, the son of an import/export agent. He was educated at Wallasey High School, then Sedbergh, in Yorkshire. He gained a first class honours degree in Engineering Science at Liverpool University, followed by a doctorate. He joined Daimler as a postgraduate pupil in 1926, gaining experience throughout the works, and formed a separate experimental department in 1929.

He installed Dr Lanchester's accelerometer for engine testing as his first task and brought the mechanism up to working state. In 1936 he was appointed technical supervisor of the design and development departments and, in 1939, assistant chief engineer. He remained in this post until 1963, with the exception of a short period at the tank design centre, near Staines, during the war. He obtained sanction to return to the Radford works, where he was largely responsible for the well known Daimler Scout Car series. With the Jaguar take over, Noel Tait was appointed group engineering consultant to the entire combine, which included Guy Motors, Henry Meadows and Coventry Climax. Interesting tasks undertaken included refining the Fottinger hydraulic coupling and adapting it to the Wilson pre-selector gearbox. He was responsible for producing the famous V8 $2\frac{1}{2}$ and $4\frac{1}{2}$ litre engines from sketches supplied by Edward Turner, the former Triumph Motor Cycle designer. He had continually to assist the bus division by solving problems of vibration, chassis stiffness and so forth. Has remained in Coventry and still runs a Daimler car.

He was a life long supporter of the I.A.E. and Chairman of the Coventry Branch, 1959–1961. Noel Tait spent a working lifetime solving problems, of which there were many, for the Daimler concern. In the early days he was almost the sole engineer in Coventry with professional qualifications. He was an outstanding scientist and a good mathematician, with a commonsense, practical approach. He worked with great energy and enthusiasm for Daimler over a period of some forty-five years, quiet, refined and unobtrusive in the manner of his kind. Such men, who seldom receive publicity, were the backbone of the British motor industry.

Captain J. S. Irving, 1880–1953

Born 29 March 1880. Educated at Coventry Technical Institute, where he gained first class honours (City and Guilds) and was also awarded the coveted Gold Medal. His first job was with the Daimler Company, which he joined in 1900, followed by a period (1902–1903) at Fulwell Engineering, of Coventry, a firm under the control of his father. He returned to Daimler, staying until 1911, although he was also teaching at the Coventry and Wolverhampton Institutes from 1908–1910, presumably on a part-time basis. He joined the R.A.E. at Farnborough late in 1911, thus being one of the original employees; this represented his first senior position, being appointed chief of engine research. He carried out work on power plants for airships as well as aeroplanes and undertook a considerable amount of flying.

With the ending of hostilities, he joined the Sunbeam Motor Company as a superintendent engineer and was appointed chief engineer in 1928. He had been responsible for development work on this company's racing and record-breaking cars as well as the advanced 3 litre sports car introduced in 1924. He also managed the racing team. He was entirely responsible for the Golden Arrow record-breaking car with which Sir Henry Segrave recorded 231 m.p.h. in 1929. This car is on exhibition at the National Motor Museum, Beaulieu, together with the previous 200 m.p.h. Sunbeam of 1927, also built under the supervision of Irving. He left Sunbeam in 1929 to join the Hunter-Hillman combine and was commissioned to design a new car with which the Rootes Brothers intended to combat the Chrysler and Buick menace in both home and export markets. This first Humber Snipe created much interest, combining good looks in a vaguely Americanised manner with a speed capability of 75 m.p.h. The car possessed an electron crankcase, a magnesium alloy then gaining a short-lived popularity, while the traditional Humber inlet-over-exhaust-valve layout was retained for a further three seasons. Captain Irving presumably felt at home in the Humber works, where he was among relatives, in particular his brother-in-law, O. D. Horton, who was service manager. Irving also took a financial interest in the Humphrey-Sandberg freewheel concern.

He joined the Bendix brake corporation in 1931 and later transferred to Girling, and was president of the I.A.E. 1936–1937. He is remembered as an intuitive engineer, always interested in high performance vehicles. He died in Birmingham on 28 March 1953.

H. C. Stevens

A time-served Napier pupil, in common with many early engineers, he then moved to Thorneycroft.

Captain Jack Irving contemplating a model of the 1927 land speed record Sunbeam, for which he was responsible

He joined the engineering staff of the Sunbeam Company in 1914, and eventually left for the U.S.A. after spending a period in the Paris office under Coatalen, joining the Oldsmobile division of General Motors. He became chief engineer, an exalted post which must have been a great contrast from the Sunbeam Company, and then spent some time as an independent consultant acting for Citroën, among others. He returned to England in about 1930 and joined the Singer Company as chief engineer and production manager. He was responsible for their six cylinder range with which they attacked the low-priced 16 h.p. market. This venture met with a limited success for a short period, but the firm lacked the resources to compete with Austin, Morris and the American imports. Stevens returned to Sunbeam in 1932 and was given a mandate to produce a new 12/14 h.p. car, probably as a final desperate measure to ensure survival.

Stevens produced a refined four cylinder engine embodying an aluminium cylinder block. The car was in the vanguard of progress with regard to suspension and appeared with a transverse spring independent layout, a scheme adopted at the same time by Delage and one or two other European makers. Stevens was not successful on this occasion and geometrical imperfections resulted in radius arm fractures. The car, complete with beautifully-made Sunbeam coachwork was, in addition, too heavy and expensive.

H. C. Stevens is remembered with affection by his colleagues as a person of refinement and culture, while it is said that to have been invited to one of Mrs Stevens's dinner parties was the hallmark of social acceptibility in Wolverhampton circles. They had a highly talented actress daughter who became one of C. B. Cochran's young ladies.

Appendix III

The relationship between Rolls-Royce and other manufacturers

A close relationship developed between Rolls-Royce, the Rover Company and Vauxhall Motors, a state of affairs brought about by the fact that the senior engineers of these concerns were all close personal friends. In particular, A. W. Robotham and Harry Grylls of Rolls-Royce were in continuous contact with the Wilks brothers and Robert Boyle at Rover, Alex Taub of Vauxhall and Maurice Olley of the Cadillac division of General Motors, the latter once serving with Rolls-Royce.

An internal memo from Rowbotham to a departmental head in response to a request from Rover for a Bentley air cleaner for experimental purposes reads thus: Do what you can to send off the cleaner quickly, they [Rover] are good friends of ours and very nice people to deal with'. A coincidence concerned the development of inlet over exhaust valve engines; both concerns were designing engines of this type during 1938–1939. There is no doubt that Rover were of great assistance to Rolls-Royce and willingly supplied information on induction systems and so forth. Rolls-Royce did, however, conceal the fact that they were seriously contemplating an attack on the Rover market with a 12–14 h.p. car using one of these new engines in four cylinder form. It is obvious from the tone of surviving correspondence that Rover had their suspicions, but they might have been contemplating a future merger of interests. Another memo from Maurice Wilks at Rover to various works foremen states: 'Show them anything they want to see', referring to an impending visit from Robotham and Grylls. There was also a request from Derby for internal dimensions of the Rover sports saloon body: 'We are constantly criticised for lack of space in the Bentley – do you suffer from the same?' writes Grylls.

Senior men from Rolls-Royce were constantly visiting various General Motors plants, and these visits were reciprocated. Frequent requests for technical help from Derby to Luton always met with a quick and friendly response. Rolls-Royce were impressed with the Vauxhall 25 and purchased one of these quite attractive big cars in 1937 for examination and technical appraisal. A letter from Robotham to Taub asks: 'How much are you paying for the 25 h.p. Frame? We think we are being fleeced for ours'.

Rolls-Royce were carefully watching the development of the new monocoque Vauxhall 10 and bought one at an early opportunity. This technically advanced car, however, did not find much favour at Derby. Robotham writes to a colleague: 'The Vauxhall 10 does not suit us, and I much prefer S. G.'s Rover, despite the lack of performance; I think this is the kind of car that we should make'.

The advice of the Humber concern was sought from time to time, and ready assistance was usually forthcoming; the relationship, however, remained formal, as it did with the Ford organisation.

A promising relationship started with S.S. Cars Ltd. Derby sought an invitation to look over the Foleshill plant of this energetic young company, and the first visit took place in July 1936. Rolls-Royce were intrigued by the ability of S.S. to build a fast, attractive and well finished sports saloon for

An experimental 14 h.p. Rolls-Royce built and tested during the war. The Park Ward body shows distinct similarities to the later Silver Dawn and Mk VI. The 1947-type bumpers have the overriders upside-down and the car appears to utilize Humber Hawk wheels and hub caps.

£385. They learnt much of interest. William Heynes, the technical man at S.S., sought and obtained help from Derby with problems of detonation which plagued his high compression engines. A further visit to Foleshill took place in November 1937 when Rolls-Royce staff were able to see the production of the latest S.S. Jaguars now fitted with all steel bodies built in house. This feature was proving very troublesome at the time of the visit, and must have reminded Grylls and his team of the problems of Park Ward, who were trying the same thing for the Bentley saloon bodies. William Lyons was then invited to Derby.

This great motor industry personality, then aged 36, must have impressed the Rolls-Royce team, for Robotham asked Hives, then chief of the Derby plant, to try and arrange to meet Lyons for lunch. 'You will find it interesting', states the memo. Rolls-Royce ordered a new 3½ litre Saloon from Lyons at the October 1937 Motor Show. The car eventually arrived, and was thrashed by the Rolls-Royce testers. An exhaust valve head broke after high speed tests at Brooklands, causing much damage to the engine. The car was returned to S.S. for overhaul, although by then out of guarantee. A charge, albeit modest, was made for this work, to which Rolls-Royce objected. The correspondence following became acrimonious and Lyons, ever careful over costs, stood his ground. Rolls-Royce eventually agreed to pay, but the promising relationship had suffered a setback.

Contact was made in varying degrees of depth with most of the American manufacturers, but apparently no real collaboration took place with European makers in the inter-war years. The one important exception was probably the adoption, by licence, of the Hispano Suiza brake servo.

No trace of any contact with the Nuffield group is recorded, save discussions with H. N. Charles of M.G. on the subject of superchargers and interest in an axle test rig built by Wolseley Motors. There is some evidence of modest co-operation with the Austin Motor Co.

Appendix IV

Rolls-Royce's surveys of competitors' Cars

A study of reports and internal memorandums between Rolls-Royce senior staff in the pre-war days is of interst. They reveal a general and serious concern over the excellence and advanced design of some competitors, particularly American, as well as giving an insight into the personal preferences and opinions of individuals.

Lanchester

A 38 h.p. car was purchased in 1912 and was driven to the Royce home in the South of France, and then back to Derby.

This make was proving serious competition to Rolls-Royce and Royce was impressed with most aspects of the design, in particular the epicyclic gearbox, which led to serious investigation of this form of transmission by his engineers.

Royce thought that the critical parts of the steering and suspension looked too slender, and thus, possibly dangerous. The meticulous Lanchester brothers would have calculated the stress of all components most carefully, and Royce need not have worried.

A cursory test of the Lanchester 40 and 21 models of the '20s sufficed. These cars were dismissed as coarse, except for the quality of ride, which was acknowledged as very good. The criticism is surprising and difficult to comprehend. The later straight 8 Lanchester does not appear to have been tried, which is equally surprising, particularly in view of the great interest in this configuration of engine at the time.

The first Lanchester built under the aegis of Daimler in 1932, the 15/18 Pre-Selector, was, however, the subject of intense interest at Derby. This car was tested by most of the Directors and senior staff during the Olympia show. This led to the purchase of one of these cars, which was exhaustively tried and investigated. It was found to be an excellent car in most respects, with a better performance than the experimental 18 h.p. Peregrine Rolls-Royce which would have shared the same market. The Lanchester would apparently exceed 70 m.p.h., and was considered a good-looking car. This is very odd; the proportions, with forward-mounted engine, were decidedly awkward, and it is difficult nowadays to enthuse over this pseudo-Lanchester.

Chrysler 72

During the '20s and early '30s it was customary for Rolls-Royce to purchase, test, strip, measure, weigh and thoroughly investigate most of the models in the Chrysler range. Rolls-Royce engineers were greatly impressed by these cars. The larger-engined models were found to have the best all-round performance of all cars seriously studied up to 1929. They were very light – partly explaining the performance – and handled well, while several innovative features were noted. Nevertheless, one does not detect any Chrysler influence in subsequent Rolls-Royce models. We learn that following on from this make, the next to find most favour in around 1930 was the Graham-Paige, which proved most effortless to drive and for which the high gearing and long stroke was held responsible.

1938: S.S. Jaguar 3½ Litre Saloon

Delivered: 16.2.38. Maximum speed: 87 m.p.h. Average fuel consumption: 15 m.p.g. Weight: 32½ cwt. Immediately returned to the makers for repairs to a leaking fuel filler neck.

This was one of the first 3½ litre cars built. Deliveries were protracted due to difficulties in the build of the new all steel body, as it was made up from badly jigged small pieces welded together. Many faults were found with this model: exhaust system fractures of pipe and mountings, door rattles, considered to over-steer badly, engine would not run slowly, and carburation was weak at certain speeds causing 'split back' on full throttle acceleration.

This car was stripped and all components weighed while the engine was subjected to bench tests. The power output was virtually identical to the 3½ litre Bentley right through the speed range. A period was noted between 62–68 m.p.h. in top gear, but no crank damper was fitted. The pedal gear and change speed lever wobbled about too much due to engine rock, while the clutch was thought too heavy to operate, and the seat cushions too small for comfort. Poor synchromexh action was in evidence – Moss gearboxes of this period were always weak in this respect, and at 6,000 miles a tooth came off one of the constant pinions, another common fault with this gearbox when asked to transmit the torque of the largest Jaguar engine.

The car was found to be an excellent starter from cold and to ride and handle well at higher speeds, although thought hard and heavy at low speeds. Nevertheless, the value for money was so good that the boys at Derby felt anxious about the competition that would result from this model when all the teething troubles were groomed out. This particular car was sold in February 1939 after 8,000 miles had been covered, and following an engine failure caused by a broken valve.

1939: Humber Super Snipe

Delivered: February 1939. Tested immediately at Brooklands. Weight 33 cwt. Maximum speed: 81 m.p.h. Fuel consumption: 13.8 m.p.g. at constant 80 m.p.h.

This car impressed with a very good all-round performance. The brakes were considered inadequate for the speed capability, and the steering too low geared, making 'driving over long distances an effort'.

The front end of the frame was found to be weak and caused 'the radiator to move up and down relative to the rest of the car as on 1–B–50 [experimental straight 8 Bentley]'. The suspension was found to be quiet and the ride met with approval: 'It takes our humped back bridge very well'. The seat cushions met with criticism. The carburettor would flood going up a 1 in 5 gradient, but was alright going down! Oil temperature reached 132°C and looked as if it might even go higher.

This car was passed on after 3,000 miles.

Essex Terraplane

The purchase of a Straight 8 Essex Terraplane in October 1933 was to prove a rude shock to the entire staff at Derby.

The silence, smoothness, performance and reliability of this cheap American seemed almost incredible, and a thorough strip and examination was authorised. It was found to be very simple in construction, a fact which only added to the general amazement. Surprisingly the steering, normally the least attractive feature of American cars, also came in for approval, and was found to be free from road reaction through the steering wheel, the most intractable of Rolls-Royce problems.

The suspension was, typically, soft to the point of allowing a dangerous degree of roll at high speeds, while the pressed steel body was stiff and quiet. The tinny appearance caused offence, of course, but Managing Director Sidgreaves remarked that 'this car is so good that it shows us up in a very poor light, and if we couldn't find anything to criticize in a car which sells for less than one third of the price of our 20/25, after import duty and delivery across the

Atlantic, then things would be really disastrous for us'.

It was discovered that the engine weight was extremely low and apparently the cast iron cylinder block had a uniform thickness of $\frac{1}{8}$ in. around the water jackets, another facet that appeared almost unbelievable. It was indicative of the state of the art of American foundry practice, pointing to a very good control of the positioning of cores apparently unattainable in England at that time, and difficult even today.

A six cylinder Terraplane was bought some months later, and this car was found to be equally impressive. The Americans were now sacrificing lightweight chassis frames in order to obtain very high torsional stiffness, and then mounting the engines in an extremely soft cushion of rubber. The weight penalty of the chassis was then more than compensated by the light engines and other chassis components.

1934: Cadillac V16

It is well known that Rolls-Royce purchased an example of one of these enormous cars in August 1934. It was imported as a special order and would easily be the most expensive competitor's car ever purchased for the purpose of investigation and test.

The car was put through the standard 10,000 mile test, carried out in France. Some trouble was experienced with the engine after 6,000 miles and the car seems to have received a mixed reception. The individual engineers at Derby and the London sales staff all seemed to form different opinions. It was generally conceded that the steering, quality of ride and high speed silence were impressive, but the car lacked refinement in other respects. It had a cheap exhaust system which suffered from surprising resonance, thus tending to spoil the general silence. The car was excessively heavy, almost scaling three tons, a factor which adversely affected acceleration, while the car was hard pressed to exceed 90 m.p.h., despite the makers' claim of a genuine 100 m.p.h.

The Rolls-Royce technical staff thought that the car had a pressed steel body, which they were anxious to study. A trip to Hoopers followed, where a door hinge was removed to reveal a normal wood body frame. The car certainly gave the impression of carrying a pressed steel body because the style exactly followed the normal Buick–Cadillac line, although stretched and widened. The body panels were in fact normal production steel parts cut in half, lengthened, and welded up as necessary. Economies of production were thus achieved and several thousand of these cars found buyers, a further indication of the enormous market for expensive cars in the U.S.A.

The 16 was, however, almost unknown in Great Britain, and the purchase of the car by Rolls-Royce would appear to have been an act of unnecessary extravagance.

1936: Cadillac V8

Major Cox, the London Sales Manager, was greatly impressed by this car. It was found to be manageable in size, handled very well and was good for 80 m.p.h.

The side valve engine was a case for some furious thinking since it was smoother and less obtrusive than that in the first Phantom III. The clutch was smooth and totally free from any trace of judder. 'It is an alarmingly good car' wrote Cox, who seems to have tried all the higher-priced General Motors offerings. He was invariably impressed by the silence, comfort and light controls which characterised these cars. The appearance of these Americans, so totally different from Rolls-Royce and other high-grade British products, also met with general approval.

1939: Buick 8 cylinder

Weight: 36 cwt. Brooklands lap speed: 91 m.p.h. Maximum oil temperature: 110°C. Average fuel consumption: 13 m.p.g.

The best feature of this car was considered to be engine smoothness and acceleration. The engine did not start easily when cold, and the valve gear was noisy. The gearbox suffered from a deal of backlash, the gear lever came too near the steering wheel and the trafficator switch was badly placed and easily turned accidentally.

'Gearnoise – good enough'. The standard of ride was thought good at low speeds, but 'dreadful when pushed hard on the road from Buxton to Ashbourne'. The car understeered, but had 'a very nice steering wheel'. Brakes were not equal to the performance on our roads, the doors did not shut easily, and the speedometer drive was noisy. Surprisingly for such a luxurious car, the seats were found to be very uncomfortable. The engine mountings were removed and tried on an experimental Bentley, but no improvement resulted; 10,000 miles were covered on this Buick.

This report was compiled by the engineering staff at Derby, who were generally more critical than the salesmen in London.

1941: Austin 8

Delivered: 3.1.41, at 2,096 miles. The first tester was favourably impressed by this small car, and considered that the stability was good, but the engine lacked performance. The comments of the second were contradictory, and the car was now stated to be heavy and generally very unpleasant due to an inefficient steering gear which was too high geared. Directional stability was now said to be very poor due to oversteering characteristics. The ride was a mixture of harshness and pitching, and it was thought that the spring ratings front to rear were badly proportioned and that Austin engineers had relied on the shock-absorber people to mitigate their poor work, which was probably right. The propeller shaft was badly out of balance and bottom gear jumped out under load. The only features to receive approbation by this tester concerned the clutch and brakes.

Rolls-Royce decided to re-build this car totally utilising their expertise and inherent refining skill as a project, coupled with their tentative study of the 8 h.p. market. New springs with carefully calculated ratings, together with a new steering gear, the latter supplied by Messrs Burmans of Kings Norton, were the main changes. A great deal of sound deadening work was undertaken, and the car was felt to be considerably improved as a result. There is no record of any consultation with Austin over this work; it was presumably confidential.

Robotham decided to instal an experimental M.40 overhead inlet engine into the Austin 8, a four cylinder type made in two sizes: 1.2 litres, and 2.1 litres. This would have been far too heavy for such a small car. The overload on the Austin transmission soon produced the inevitable breakages, and the increase in weight over the front axle necessitating yet more new springs; it must have ruined the steering and upset the brake front-to-rear effort. This work would not prove much of value.

The sanction of such activity when war work must have been at its most urgent pitch in the summer of 1941 is very difficult to comprehend. Robotham used the car himself in this form for some 2,000 miles, when the Austin engine was re-installed and another B.40 engine was inserted into an Austin 10 of alligator bonnet design similar to the 8 h.p. These two cars were later passed over to the works transport pool for general hack use.

Citroën

The front wheel drive Citroën, introduced in 1934, naturally excited some attention and Walter Sleator at the Paris depot was commanded to obtain a complete set of suspension and drive shaft pieces of this type. The assistance of Citroën Director, Pierre Michelin, was forthcoming and the parts were duly delivered. Grylls strongly recommended the purchase of a complete long chassis version of this car but this does not appear to have materialized.

1935: Hotchkiss

A product of the old established armaments firm from St Denis. The rather plain appearance of Hotchkiss cars was deceptive, concealing high grade engineering and workmanship together with an outstanding performance in the case of the later models. The following Rolls-Royce internal correspondence, proves the point.

To Sg. from Hs/Rm.   Hs/Rm.13/KW.24.4.35
c. to Wor.
c. to E.
c. to By.
c.to Bn.
c. to Bly.

<u>3½ Litre Hotchkiss</u>

This car has a very outstanding performance. To give comparative figures of the Autocar test between the

Hotchkiss of April 12th, and the Bentley which was taken 12 months ago. The Bentley then weighed about 1½ cwts. less than it does now.

For actual top speed the Hotchkiss is slightly better than the best conventional Bentley we have ever made. The acceleration is of course a great deal better than the Bentley when the gears are used, but not below 30 m.p.h. on top gear, because the top gear ration of the Hotchkiss is 3.6 to 1 against our present standard of 4.1. to 1. The reason for the acceleration of the car – apart from the fact that the engine power output must be quite good – is its low weight, as it is only 7 lbs. over 30 cwts.

Mechanically, the valve gear is quieter than the Bentley. The absence of an intake silencer, and a noisy exhaust system, give the car a feeling of courseness which we do not think it really merits. The expenditure of a trifling sum on adequate silencers would improve the car out of all recognition.

The steering does not lack selectivity, but again is not altogether pleasing to handle. It seems that there is nothing fundamental about this, but that it would be completely corrected by the latest type of double toothed Marles.

The suspension is not at all bad from the sports car point of view.

The brakes are most effective at ordinary speeds, but we understand are of the Bendix type, which from our experience might be expected to be inadequate if the car was driven to its limit on winding roads.

The car tested was the two-door Saloon illustrated in the Autocar of April 5th.

Summarising the impressions obtained, there is no doubt that this car can out-perform the Bentley as at present being sold, to a somewhat depressing extent, particularly for acceleration but also for maximum speed. Its faults seem largely unnecessary, and if it was taken in hand by anybody with knowledge, it might easily be made into a nice motor car. The fact that it has a 5" shorter wheelbase than the Bentley will affect the coachwork which could be fitted.

We think that this is yet another indication that we must improve the Bentley performance.

AUTOCAR – ACCELERATION FIGURES 20–40 M.P.H.

	Top		Third		Second	
Bentley	Hotchkiss	Bentley	Hotchkiss	Bentley	Hotchkiss
8.8	9.8	7.0	5.4	5.2	3.8
8.8	9.6	7.0	5.6	5.2	4.2
9.2	9.4	7.4	5.6	6.0	

Through Gears

	Rest to 50	Rest to 60	Rest to 70
Bentley	13.2/5	20.2/5	
Hotchkiss	10.4/5	13.4/5	20.4/5

Best Half Mile

Bentley 91.84
Hotchkiss 95.74

Our figures – Lap Speed

Experimental 3½ litre

1–IV Bentley   *Hotchkiss*
Closed 87.0   87.84
Open   88.78   88.94

Hotchkiss
10–30   8.375
10–60   24.5
10–60 Gears   13.5
 0–80 Gears   33.3

1937: Lagonda V12

W. O. Bentley found he had to complete his service contract with Bentley Motors Ltd after the Rolls-Royce takeover. He suffered considerable humiliation but nevertheless made good friends with some of the engineering staff, even persuading Stuart Tresilian to join him later on as chief designer at Lagonda, to which Bentley went as technical director upon his release from Rolls-Royce in 1935.

The following correspondence within and from Rolls-Royce relates to the recently-introduced twelve cylinder car which was prevented from seriously affecting Rolls-Royce sales by the limited resources of Lagonda and the outbreak of war.

H.   COPY
c. to Cx.   AA2/BK24.12.37

<u>Re the 12 cylinder Lagonda</u>

At lunch time on Tuesday last I was fortunate enough to be able to have a run on the 10 ft 4″ wheelbase chassis Lagonda fitted with the Touring Saloon body.

I suppose I did about 25 miles on it and one's first impression was remarkable smoothness right through the range, and ease of control. The latter feeling was a little upset as one went on by a rather heavy synchro-mesh action in the gearbox. I say heavy when comparing it with our $4\frac{1}{4}$ Bentley box. Possibly to many people not used to a box of such refinement this heaviness would not have been noticeable or unpleasant. Another thing which also helped to remove the feeling of lightness was the feeling of pushing oil through pipes when the foot brake was applied. This of course one would get used to, because it is, I believe, a feeling always associated with hydraulic brake systems.

The smoothness of the engine is undoubtedly remarkable and it is also extraordinarily silent, although the engine gears could be heard on the particular car I tried. Whether this is characteristic of the engine or not I cannot say, as there was no other car with which I could compare it. If this gear noise was overcome the engine would be every bit as quiet as the $4\frac{1}{4}$ Bentley.

The performance at low speed in top gear was not quite what one would expect. It is nothing like as good as the $4\frac{1}{4}$ Bentley. This could undoubtedly be overcome by lowering the axle ratio which could be done without detracting in any way from the 3rd gear performance of the car which is remarkably good, as it attains 90 m.p.h. in this gear. If the axle ratio was lowered to a maximum of 80 m.p.h. in this gear the low speed performance in top gear would, I believe, be vastly improved. I have no doubt that "W.O." will make this modification, which will make the chassis a very fine performer.

Another thing I noticed was when I was cornering fast and I rather deliberately put the car into the bad road surface at the side of the road to see how it behaved. Its front suspension is undoubtedly very good, but I felt that the back springs were too light, and the shock absorbers had not got sufficient control. Mr. Tresilian who was with me agreed, and said that they had actually been carrying out some experiments on these dampers which he did not think were up to standard.

My impression of the run was that it was a very pleasant car and I much enjoyed myself, but I felt that personally one would get tired of the heavy synchromesh gear box and the slight engine gear note which was audible from the engine at most speeds. If both these points were improved and one got used to the feeling of hydraulic brakes I think one would have a very fine car indeed.

I gather they have had a certain amount of brake judder with the Lockhead [sic] brakes, which they fit, which was noticeable at high speed. The car which I tried had had certain modifications carried out to overcome this trouble, and it certainly did not show up when I was driving.

I think I have tried to be quite unbiassed in forming my opinion of the car, and I think that with a very small amount of work this chassis is undoubtedly going to be one that will have to be reckoned with as a competitor of the $4\frac{1}{4}$ Bentley.

Perhaps I should point out that when I talk about low speed performance I merely mean acceleration from low speeds in top gear and not unpleasant rattles and snatches in the transmission. This latter is remarkably smooth and free from rattles.

                              AA.

Rm/Gry4/R.   15th June, 1938.

W. O. Bentley, Esq.,
Lagonda Motors Ltd.,
STAINES,
Middlesex.

Dear W. O.,

Isn't it about time I tried one of your 12-cyl. motors? I shall be in London on Wednesday next week, and wonder if I came to Staines if you would have a car available.

Rm. also much wants to try one, but being very honest insists of my telling you that if asked afterwards will, of course, say ours is a better car. On the other hand, when we have something new on show, you shall certainly have a run on it.

No one will be offended if your reply to this letter is not couched in parliamentary terms – merely disappointed.

Should this meet with your approval, and we arrived about mid-day, will you lunch with us somewhere suitable for car testing?

Your sincerely,

By. from Rm.
c. SG.

c. Hs.   Rm10/R.23.6.38.

12-CYLINDER LAGONDA.

W. O. Bentley was good enough to let us try his own 12-cylinder Lagonda car on Wednesday, and also let us examine the pieces in process of manufacture.

In addition to an ordinary road test we took the car on Brooklands. The general impression we got from the car was that it represented a distinct achievement having regard to the time and resources at the disposal of the Lagonda Co. It is evident that Trisillian [sic] put in a great deal of hard work in designing this car.

The intermediate length of chassis gives a body space which very closely approximates to that available on the Continental P.III. Without taking accurate measurements we should say that the Lagonda has slightly more room. We were told that the complete weight of the car we tried weighed just under two tons. Considering the accommodation it is a very meritorious figure, a similar P.III weighing 52 cwts, or more.

The most attractive feature of the chassis is, undoubtedly, the power plant, which is remarkably smooth, and, from the Brooklands figures must have a very good top power output. Up to 40 m.p.h., however, it is somewhat flat, and, for this reason would not be a nice car to drive in top gear in London traffic.

From the point of view of the occupants of the car, it approaches the Rolls-Royce standard silence at all speeds, except for tyre groans and an exhaust boom round about 30 m.p.h.

The steering is too heavy at low speeds, but in view of the weight on the front tyres, this cannot be fundamental, and it will not take them long to put it right. It is, apparently, due to the poor design and inferior workmanship put into the Marles steering which they fit.

The gearbox is reasonably quiet, but the big central change-lever spoils the freedom of access to the front seats.

At the moment they are suffering from heat in the front seats. Here again, it is only a matter of knowing how to put it right and the trouble can be eliminated.

The brakes are not adequate to deal with the high top speeds attained by the car. They will probably deal with this by fitting some sort of servo.

The directional stability of the car at high speeds is as good as that of B.III, but it is hard work driving the car at low speeds round lanes.

We understood from W. O. that they were producing about 5 cars per week, and had already sold about 70. Their weakness would seem to be that they have lost their best men just when needing them most, i.e., when service troubles are beginning to appear. These troubles seem to be rather overwhelming W. O. at present.

At Brooklands the car achieved a lap speed of 93.5 m.p.h., and we understand that the short chassis has done considerably better than this. A recent figure obtained on 6.B.IV on the overdrive was 89 m.p.h.

Summarising the impression created by this Lagonda, we think that B.III would be infinitely more desirable to drive as a Sports car, but that we shall have to try and get a top speed performance comparable with that of the Lagonda to justify our smaller seating accommodation. The P.III, or the Wraith, are, of course, much more desirable town carriages. For a man, however, who desires a combination of both types of car, the Lagonda will represent a fair compromise when they have got their steering, brakes and durability up to the required standard. The success of the Lagonda enterprise, however, would seem to depend almost entirely on whether they can deal with, and adequately survive the complaints which are bound to arise during the first 12 months the chassis is in the customers' hands.

At the moments they appear to have more than enough aircraft work to compensate for any falling off in car sales.

Rml/R.   23rd June, 1938.

W. O. Bentley, Esq.,
Lagonda Motors Ltd.,
STAINES,
Middlesex.

Dear W. O.,

I should like to thank you for the very pleasant time you gave us yesterday.

Seeing your house and garden was an unexpected pleasure.

With regard to the car, I can honestly say that I am astonished at the standard which you have reached in so short a time, particularly in view of the fact that you have had to educate the Factory with regard to manufacturing.

The engine in particular struck me as being a very sweet running unit.

I hope that you will not be overwhelmed with Service troubles, which are always the worst part of any new motor-car.

We are looking forward to having the Lagonda as a competitor, which will prevent us sitting back and having a rest. As soon as any of our recent experimental stuff reaches the public, we shall be glad to give you an opportunity to trying it, if you are interested.

Your sincerely,

Appendix V

Internal Rolls-Royce correspondence about Bugatti

Memo from Henry Royce to Hives, 2 April 1928.
Written from France.

Re. Steering

The car which (after the Hispano) might be interesting from a road holding point of view, and for other reasons, is the straight eight Bugatti.

I see several of these 1927 models for sale secondhand at about £400. We should not drop much in buying and selling one. I was thinking of getting one myself just to study, but perhaps I am too big, and certainly I am too old: it might appear rather crazy.

I say this because these cars, above all others, are used on these rough and varying roads with much confidence that they will not get out of control, and it might settle some of the questions about which there are different opinions, but in all probability what suits them would not suit us, but it would be as well to find their characteristics.

Memo from Hives to Paris office, 4 April 1928

We are thinking of purchasing a straight-8 Bugatti. It is not necessary for us to have a new car; a second-hand one in good condition would answer our purpose.

Will you please let us know the approximate cost of such a car.

Memo to Royce from Hives, 25 April 1928

8-Cyl: 2-Litre Bugatti
We had a run in one of these cars yesterday. The owner wishes to sell it for £450.

The general running of the car was impossible. The springs were solid, the carburation was very tricky; if the accelerator pedal was pressed too far, the engine stopped, or else it was liable to catch fire.

The steering of the car was perhaps the best feature on it, but it is only what one would expect with a very light car, small wheels and tyres, solid springs etc. The car is so far removed from any of our products that we doubt whether we should gain any useful information.

When one actually tries these cars which are built for speed, such as the Bugatti, the Lorraine Dietrich and the Mercedes, and experiences the atrocious riding at low and moderate speeds, one considers our effort on the Phantom Sports car was an achievement. We are certain there is no car in the world with the speed range, flexibility, silence and comfort equal to the Phantom Sports car.

Memo from Paris office to Hives, 26 April 1928

I enclose price-list of standard Bugatti cars, and also a list of four second-hand cars showing prices, these latter being of the straight-eight type, but you perhaps require to get hold of one of the big models, of which only very few have been turned out. This may be rather difficult, but I am having enquiries made, and will let you know as soon as I have any further news on the subject.

In the course of a conversation with Mr Bugatti some time back, he offered me one of his new big cars for a trial any time I liked. So far, I have not availed myself of Mr Bugatti's offer, but if you so desire, I could arrange to have this car as soon as I know when you are likely to be in Paris so that you could try it with me.

However, in accordance with information from outside sources, this car does not appear to be a great success, and a good many alterations are likely to be made before this car is put on the market as a standard type.

Memo from Paris sales to Hives, 4 September 1929

I have been given very reliable information that the Bugatti 5-litre 8 cylinder chassis has given very satisfactory results. It is intended to compete with the Hispano type of car, and is capable of 150 K.P.H. with an inside drive body. As a chassis, speed of 165 K.P.H. is obtained.

The 16 cylinder or double eight will not appear on the market at least for a further considerable period, but a decided attempt will be made with the 5 litre eight in line to capture our class of trade.

[In pen at the bottom of this note is written 'Best of luck for the Schneider Race, I shall be at the works on September 16th' – the luck held!]

Memo from Hives to Works Manager, Wormold, 16 September 1929

One of our men Tresilian has an eight cylinder Bugatti car. [Type 35 racing model].

At various times R. [Royce] has been interested in the construction of this engine especially in relation to the proportions of the crankshaft, etc. At the present time one of the big end bearings have [sic] gone wrong and it is necessary for the engine to be dismantled and we recommend that we dismantle the engine in the Expl. shop. Tresilian to pay for any repairs which are made but R.R. will stand the cost of dismantling and erecting and that we shall take any particulars of the engine that we require.

Memo from Tresillian to A. Elliot [Chief Engineer] 2 June 1930

Re. 8 Cylinder Bugatti Engine

Mr Jenner asked me to send on prints of the crankshaft, etc, of my car, since you are now interested in straight eights. I thought a few remarks would also be of interest:

1   *General*   The engine is 60 × 88 m.m., two litres capacity; compression ratio approximately $6\frac{1}{2}/1$. Considering its power and performance it is very flexible indeed. There are two inlet valves and one exhaust per cylinder. The timing is:

   Inlet V.O.: T.D.C.   V.C.: 20° A.B.D.C.

   Exhaust V.O.: 50° *BBDC*   V.C.: 12° A.T.D.C.

   The exhaust valve timing is fairly 'racing' but, with the ample inlet valving and ports the inlet time is very 'touring'. I believe this is one secret of the remarkable combination of power and flexibility i.e. they get the power without the aid of an extreme inlet timing which would spoil the low speed performance. Top gear is 3.84/1. Petrol consumption about 18 M.P.G.

2   *Crank Arrangement*   As shown in the small diagram on the larger print, this differs from that of any other straight eight. The actual firing order is 15263748. This order is at least easy to remember since the two halves fire in numerical order alternately. There are two Solex carburettors, each feeding into a two-branched induction pipe with a small water jacket, each branch feeding two cylinders via a common inlet port. This port contains four valves (2 per cylinder).

   The crank arrangement does not appear to help the distribution, so I think it must be chosen to facilitate the design of the built up shaft.

   There are five main bearings, three large double-row self-aligning Skefkos, and two intermediate roller bearings, having split outer races. Another advantage of the crank arrangement is that there are only three main bearings highly loaded at high speeds, and the inertia loads on the shaft are more balanced than in the usual arrangements. This should help to give smoother running. The normal 2–4–2 or 4–4 arrangements would have resultant inertia loads on all five bearings. It would appear a good arrangement for a high speed engine. The maximum permissible R.P.M. is 5,500, at which speed the car is geared to do about 120 M.P.H. A previous owner is reputed to have captured the 1 Hour record at Brooklands with it at 117.

   The shaft was very tight when we took it down, but had been fretting slightly on most of the taper wedges etc, presumably during C.T.V. [crankshaft torsional vibration] periods.

   The wedge-shaped dowels that grip the crank-pin extensions are not actually flat as shown in my drawing, but very slightly radiused lengthwise (and also hardened) so that they bed definitely at either end of the flat on the crankpin extension. [Not correct]

3   *Balance and General Smoothness*   I have worked out the balance as follows – [we have omitted the technical analysis]

   There is a distinct vibration period producing movment of the whole car, at about 1200 R.P.M. running light, but it cannot be noticed under load. There is no other vibration period in the speed range which could be blamed to out-of balance. The engine is uniformly smooth, up to maximum R.P.M.

   The over-run is very smooth. It has four solid engine feet, on the bottom half, and the crankcase is evidently intended to stiffen or steady the front of the frame. The ratio of rotating weight to reciprocating weight per crank is about 10 times what we are used to, and the crankcase is extremely stiff, both halves.

   The car I have is known as the 'Full Grand Prix' type; there is a plain-bearing edition of it known as the 'Modified Grand Prix', with only three main bearings, which is astonishingly rough, and hardly recognisable as an 8 cylinder engine at all.

   C.T.V.   There is a main period at 4500 R.P.M.

and a very slight period under power at 2250. No others are detectable. No damper is fitted. The R.P.M. of these two periods has not been affected by stripping the shaft and reassembling it again. The flywheel is really only a casing for the multiplate (steel and cast-iron) clutch, and is only about 9″ diameter, and very light. The node would probaly be about a third of the way up the shaft.

Big End Bearings I had one of these pack up, which is the reason why we pulled the engine to bits. The firm very kindly paid the cost of stripping and reassembling of the crankshaft and crankcase, and I paid for the new bits, made by RHC [Coverley].

When examined, it appeared that the rollers (Hyatt type) had broken up, and smashed the bronze cage as well. Three other big ends had one or more broken rollers. The rollers appeared to be of a very weak design, and we made a complete new set merely hollow, i.e. omitting the spiral groove, together with C.H.N.S. cages lightly hardened instead of bronze. The big end eyes are very stiff, and evidently intended to remain circular. Lubrication is via the oil jets shown. We have subsequently come to the conclusion that these rather fragile looking rollers are put there intentionally, to be the weakest link, in the same way that we use white metal. In the event of trouble (e.g. oil shortage) they would crumble up, and save a broken rod which might wreck the engine, or avoid straining the expensive crankshaft and case. It is not so difficult to 're-metal' the big ends occasionally by cleaning up the pins and rods and fitting oversize rollers.

I hope to be able to send you further information about the car when available (as various bits pack up).

Appendix VI

Correspondence between Dr Lanchester and Percy Martin

Projected new post-First World War design by Dr F. W. Lanchester outlined in a letter to Percy Martin, Managing Director of The Daimler Co., a typical case of advanced thinking by Dr Lanchester when at the height of his powers of innovation. The scheme was not supported by the Daimler or Lanchester firms, who clearly saw that it was far too advanced for 1919 and would, in addition, require an enormous amount of costly development work.

F. W LANCHESTER.M.INST.C.E.
                              41, BEDFORD SQ
                                   W.C.1.

Percy Martin Esq.,   October 15th 1918
  "The Spring"   St. David's Hotel,
    KENILWORTH   HARLECH

Dear Martin,

    I am addressing this to "The Spring" as I have not your present address, but doubtless it will be forwarded.

    The subject I am dealing with is the design I have in hand here embodying the hydraulic features. Firstly I will give you an outline of what I am doing:-

SUSPENSION

    Steel springs are being entirely done away with, also transverse axles; each wheel is being controlled by its own individual mounting which consists of a single vertical guide pillar in the case of the front wheels, which also serves as steering pivot, and in dual guide pillars in the case of the rear wheels, one a few inches forward and the other a few inches behind the axle line. Both steering and driving wheels are being splayed (according to our customary practice in the case of the front-wheels is, as now, the reduction of the distance from the point of road contact and the pivot axis. in the case of the back wheels the advantage is fourfold:

    (1) The point of road contact is brought more directly under the hydraulic support.
    (2) For the accepted standard of wheel gauge, 5-inches of extra rear body width is available.
    (3) A better clearance is given to the hydraulic replenishers.
    (4) The clearance under the worm box is increased $1\frac{1}{2}$-inches (this point will be better understood in the light of further description).

    The worm transmission box is carried directly on the chassis and thereby becomes suspended instead of unsuspended weight and the drive of the rear wheels is by a jointed shaft in a similar manner to that used some time back by the De Dion Company. There is a novelty introduced in carrying out this drive, a novelty in fact that renders it possible, in that the outer universal joint is carried on the hub cap, the jointed shaft being carried through a hub of large diameter, the shaft is thus a foot longer than would otherwise be possible; the actual length of the coupling shafts is 22-inches, which I believe is adequate.

    The hydraulic plungers are contained axially in the guide pillars and the heads of the plungers bear on the blind ends of the guide sleeves; each rear wheel thus has two hydraulic plungers and each front wheel has one.

    The system of maintaining the oil pressure is that covered by my patents and consists of a replenisher which automatically comes into operation when the suspension gets down below a certain point. All that is necessary for the attendant to do is to see that the oil supply reservoir, containing about a pint, is duly filled when required.

    The clearing away of all moving under-axles and axle work, torque tubes etc. results in an enormous simplification of the whole design and, presuming everything comes off all right, will I think be considered a very great advance. It allows of the whole space between the two chassis members being utilised and I have arranged for a spare wheel locker under the floor of the rear body and a 20-gallon petrol tank amidships, the top of which only projects a couple of inches above the chassis members. There is no longer anything requiring access in the body work and no need to arrange for a hinged or lifting body.

POWER UNIT

The hydraulic clutch gear box, which is one of the new patented features, consists virtually of a drum shaped fly wheel bolted direct to the crankshaft containing the whole three speed and reverse change mechanism. It is housed in an aluminium case bolted to a face on the rear end of the crank case and base and has the control lever, by which all clutching and change speed is effected, directly connected thereto. Thus the car will have a central control after the manner of some of the American vehicles, but if this is objected to the control lever can be carried to the right hand of the driver. The transmission from the power unit to the worm box consequently becomes a straight tubular shaft about 7-feet in length and runs close under the petrol tank and spare wheel locker.

BRAKES

Direct road wheel axle brakes are fitted, but instead of being fitted to the road wheels themselves they are arranged one on each side of the worm gear box, the weight of the brakes consequently being taken off the unsuspended part. This disposition had other advantages of a more important kind. The brakes as above are controlled by two rods running longitudinally one on each side of the propeller shaft and are worked direct from a foot pedal. There is a hand brake lever fitted in the usual place also operating the same brake in order to hold same on when the car is left standing. The hydraulic gear itself can be used as a brake and it is contemplated that in the ordinary way it will be considerably used in this manner. Should occasion demand it, however, an additional brake may be provided and be applied by either the hand lever or the foot pedal, in place of both operating the same brakes, as in the experimental design in preparation.

UNDER-CLEARANCES ETC.

The new design is coming out wonderfully clean beneath. The worm is arranged under the axle and the clearance is about normal, that is to say between ten and eleven inches, this is in fact better than in many existing designs owing to the fact that the rake of the rear wheels gives an inclination to the mean position of the axles and so the centre line of the worm wheel is about 1½-inches above what it would be in an ordinary design (this is the fourth point mentioned in giving the advantages of the splayed rear wheels); it is desirable to have this additional clearance in view of the fact that the worm box is now on the suspended portion of the vehicle and rises and falls with same. Practically speaking the underneath of the car is a flat, smooth surface only broken by the protruding base chamber of the engine and the under half of the hydraulic gear box. This combination is of smooth stream line form. The exhaust box, which is being arranged on the inner side of the car directly under the chassis member, and is a cylinder 6½-inches diameter × about 4-feet in length. The floor level is approximately 24-inches but will vary with the state of the suspension. Thus, after long standing it would sit down somewhat which is actually no disadvantage.

CLEANLINESS OF DESIGN ETC.

The whole design lends itself to a general cleanliness and absence of rods, shackles, lugs etc. The chassis members are almost straight and narrowed at the front necessary for the steering lock; this is given by a uniform taper of the chassis from front to rear, the width of the chassis at the bonnet being 30-inches and at the rear axle line 42-inches. torsional stiffness is given to the chassis by four transverse tubes. One of these tubes constitute the rear element of the chassis and at the same time form the hydraulic reservoirs of the rear suspension and carry the pillars on which the rear wheel axles are mounted. Incidentally also these rear tubes carry at their middle point the worm box. The fourth tube is situated about half the length of the chassis just in front of the petrol tank and serves incidentally for the attachment or a transverse way shaft by which the brake motion is conveyed from the pedal gear to the longitudian actuating links.

With a view to more completely clearing up the obnoxious details with which cars of the present day are encumbered I have (in course of design) a combined instrument and switch board to be situated just beneath the steering wheel. I have designed a wheel on the lines used by the Humber Company with a single tubular spoke. If this is well made it will not look amiss and it has several important advantages. The whole of the internal mechanism of the steering pillar becomes external mechanism. The steering pillar itself consists of a 4-inch tube through which the steering shaft passes, and in the space between the steering shaft and the pillar I am accommodating the various way-shafts to control the ignition timing, the throttle, the jet, the dual ignition and the conduits for the electrical connections of the switchboard and ammeter and voltmeter, also the shaft to drive the speedometer. The various functional connections can be made at the lower end of this pillar at any convenient points, they do not have to be carried through the middle of the steering shaft as a present and the whole design makes a clean sweep of all the objectionable features with which we are only too well acquainted. The fixed dash with its instruments in all sorts of inaccessible and invisible positions disappears. In the detailed carrying out of this scheme I am designing the "instrument head" as we may call it, in three pieces. There is one piece which I may call the Daimler section which includes and contains the throttle, jet and ignition levers, the electrical conduits and the speedometer driving shaft. There is the electrician's section, which is an independent unit secured to the Daimler section by a couple of screws, all electrical joints etc. being made on the plug system when this unit is put into position; there is lastly the speedometer unit which likewise is put on as a separate piece by a couple of screws and which contains in a straight line row the speed indicator, the mileage recorded, the trip indicator and a timepiece. The whole is being carefully studied with a view to making it weather proof and the idea is that the two separate components may be manufactured to our specification and block drawing by a speedometer firm and an electrical firm, ready to put on to the Daimler section without anything more than a couple of screws (or bolts) and unskilled labour.

The key to this scheme, which I think you will consider very attractive, is actually the one spoke steering wheel, since in order to carry the idea into effect the head fitting has to be under the wheel instead of above it. Incidentally, a bad feature that has crept in owing to the mass of gear internal to the steering shaft is being done away with, namely the worm steering transmission is able to be designed for its own sake and the worm does not have to be made of undesirable great diameter to include the various four or five concentric elements which have at present to come within it. My own opinion is that the 4-inch diameter steering pillar with the instrument head will look a strikingly attractive feature and will give a sense of strength and robust construction that will make the ordinary present day construction look like "back numbers".

Admittedly the whole scheme contains so many novelties that it must be looked upon as including a large element of gamble, but I think you will agree that it is a legitimate one in view of the fact that no one feature definitely depends upon the other features. If we could win all along the line the success would be gigantic; it would pulverise people like Rolls-Royce and would paralyse the components business as it has developed in the United States. It is really an immense stake to play for.

Within a fortnight or three weeks I shall want to get much more drawing office strength on to the job, in order to get the design to such a stage that questions of weight and cost can be accurately worked out and our position with regard to both these vital points can be ascertained. I think that will be the occasion to bring the work down to Coventry and I hope that I shall be able to get at least two or three men to help out of the existing drawing office staff. I think I could make use of Balcombe (an electrical man) who is not as far as I known swinging his weight at the moment; I should like to get Dixon on to it also and one of the juniors.

I am writing you this full report as to what I am doing and my intentions, so that you will be able to chew it over and be prepared for a discussion when I next see you. I shall then be in a position to put some of the leading drawings before you, although owing to the fact that there is hardly an existing feature of the present day motor car in the new design that has not undergone substantial modification, it will yet be some time, even with the assistance required, before complete drawings can be ready. One has to go through the whole design in the rough and then go through it all again making revisions in a little greater detail and then still once more in order to shake down the various points of novelty into a good homogeneous design, but it is getting more and more satisfactory with every revision and the real difficulties have now been all eliminated.

I think you will agree that my estimate of the end of the War as being due next year looks like coming off; I do not attach a great deal of importance to the present peace negotiations, because broadly stated, the Hun has got to come off his perch and give an unconditional surrender, that is the only way I can see a finish and I think it may easily take another six or twelve months to bring that state of things about. However there is the possibility of an early collapse and this gives considerable point to my suggestions relating to the Rocket development. To some extent the two jobs want considering together as they are competitors in the same labour market. The supply of competent draughtsmen is limited.

Please give my kind regards to Mrs. Martin. I hope the change of air is doing you both good. I feel inclined to get on the train and come and have a few hours' talk and breathe a little of the stuff myself.

Yours very truly,
F. W. LANCHESTER

P.S.   I have just been looking at the Timken Magazine. I do not like counting my chickens before they are hatched, but I cannot help thinking how it would put the fear of God into them if an entirely new system of construction were to come into vogue!

P.P.S.   I have just managed to secure a small quantity (122) Havana cigars – will enjoy one or two with you when we meet.

F. W. LANCHESTER, M.INST.C.E.
                    41 BEDFORD SQUARE,
                            W.C.1

                       October 30th. 1918

Percy Martin Esq.,
  "The Spring"
    KENILWORTH

Dear Martin,

    I duly received wire last night saying that you are indisposed and cancelling my visit to Coventry today. I have this morning confirmation of same in which Longmore tells me that you have taken a chill. I trust it is not an onset of an attack of the 'flu'. At any rate it is some comfort to know that Mrs. Martin will insist on you taking care of yourself. I am sure she will not let you take any risks.

    I am hoping this week to clear up the design of the proposed experimental chassis – I am at work on some of the final details. After that it is a mere matter of getting tracings etc. and one or two sets of prints, so that we can lay the scheme out on the table for discussion and also look into the matter of costs, weights etc. I expect to boil the whole design down into a set of twelve or fourteen key drawings from which the details can be subsequently prepared; these are now nearing completion – another ten days I think will clear them up (probably too sanguine – say a fortnight).

    The next thing to tackle is the engine. I cannot make up my mind yet whether the American practice is right in eliminating the magneto. You will remember that we were just in this vein of thought when the War broke out and I do not know that the experience or evidence that has accumulated since has been decisive one way or the other. I believe the American practice in this respect is to embody in the lighting-set dynamo a contact breaker and distributor geared down to the required speed so as to avoid one of the driving connections to the engine.

    I rather presume that the forced lubrication experimentally adapted to the 30-h.p. will be adopted as our standard. Apart from the lubricating pump there have been altogether too many gear shafts etc. to provide, namely magneto, circulation pump, fan, lighting dynamo, starting motor and, in the cases where dual ignition is fitted independently of the magneto, a separate distributor shaft. Beyond these we may include the starting handle, making that is to say six in all without counting the separate dual. I have included the starting handle since it normally occupies the end of the crankshaft, which might otherwise be used as a direct means of driving one of the necessary fitments. For example in a little experimental four cylinder that we had on test just before the War we had arranged the magneto to be driven from the tail end of the crankshaft direct and, owing to the small size of this engine, we had fitted the starting handle to the eccentric shaft.

    I am inclined to adopt the scheme of using the crankshaft end for driving the lighting dynamo direct and arrange the starting handle on gear a little to one side, gearing down on to the engine, so that the starting handle becomes more or less a "barring" round contrivance, possibly two revolutions of the starting handle to one of the engine. If we can do this and if we can embody a contact breaker and distributor in the dynamo itself, and assuming that the fan is driven by a belt off the damper, the number of extraneous driving connections required is reduced to two, namely the circulation pump and the starting dynamo. I think this as a scheme is worth going for, but there is a good deal of work to be done.

    The starting dynamo lends itself to a neat arrangement in my new scheme, the usual plan of a gear driving on to the fly wheel being adopted; the circulation pump could be driven by a spiral gear off the cam shaft, as in our old 38. The spiral gear in the old 38 did not stand up any too well, but I think it would be all right relieved of the magneto.

    If possible I want to return to the practice of putting the chain drive on the front end of the engine. It is badly in the way in the rear position and interferes with the gear until being bolted to the engine as is necessary in my experimental scheme. It can be managed experimentally but not in a manner that would be fit to perpetuate.

    The above are some of the points that I want to ventilate and discuss with the Technical staff when I put the new scheme up.

    I rather gather that the Lanchester Company would be open to associate themselves in this experimental work if you see fit to entertain a suggestion in that direction. It might be made an occasion to bargain for them to drop the worm manufacture and turn their plant over to us; the idea just passes through my mind as perhaps worth trying. I think we came to the conclusion that in introducing anything strikingly novel it would be more effective if we could do it by a small group of Companies under Daimler leadership, than ploughing a lone furrow, as at first in the case of the Knight Engine. Assuming success, and I always assume success, we should have to convince the public that the new thing is right, and the public take a lot of convincing if the rest of the trade (in self-protection) form what amounts to a conspiracy against us. It might be easily convinced if three or four of the leading firms were to all strike at once. This I put up as a subject inviting your views.

    Hoping you will speedily be restored to health, and with kind regard,

                      Yours very sincerely,
                      [F. W. Lanchester]

Index

A
A.40 (Austin) 96
A.E.C. 50, 53
Acres, F. A. S. 40, 43
Ainsworth, Harry 99
Albemarle St 70
Alfa-Romeo 72
Alsace 75
Aluminium Corporation 54
Alvis 20, 64, 76, 93, 134
Armstrong-Siddeley 111, 161
art deco 70
Ashbourne 176
Associated Daimler 50
Atalante 175
Auburn-Cord-Dusenberg 90
Austin 65, 111, 170, 172
Austro-Daimler 166
Autocar 59, 176, 177

B
B.S.A. 76, 107, 134, 150, 161
B.S.F. 103, 145, 146
Baker & Perkins 158
Balcombe 187
Ballot 64
Barcelona 64
Barclay, Jack 80, 110
Barnsley, Hamilton 25
Bastow, Donald 160, 161, 162
Bastow, Geoffrey 161
Bavaria 165
Beaulieu 169
Beauvais, C. P. 43
Bendix 43, 169, 177
Bénès, President 166

Bentley, W. O. 161, 162, 178
Benz 40, 166
Bertrand, Edward 165
Beverley 90
Birmingham 27, 169
Birmingham Corporation 53
Birmingham Town Hall 20
Black, Sir John 153, 168
Bohemia 166
Boissière, Rue 165
Bordeaux 165, 166
Borg & Beck 86, 150
Borg Warner 111
Bosch 78
Boyle, Robert 171
Bristol Cars 162
Bromley 125
Brooklands 158, 172, 174, 179
Brown, David 161
Bugatti, Ettore 76, 146, 164, 165, 166, 181
Bugatti, Jean 75, 78, 164
Buick 90, 93, 102, 169, 175
Burmans Ltd 176
Bush, Lt. A. E. 50
Buxton 176

C
Cadillac 90, 171
Canada 93
Cappa 165
Charles, H. N. 172
Chesford Brook 89
Chausson 78
Chevrolet 89, 90
Chrysler 43, 90, 127, 169, 173
Citroen 170

Clayton-Dewandre 57
Clipper, Packard 90
Coatalen, Louis 28, 150, 151, 170
Cochran, C. B. 170
Colmar 78
Cologne 164
Colombo 166
Cord 90
Corsica 80
Cotal 78
Coventry 26, 45, 57, 169, 187, 188
Coventry Climax 169
Cox, Major 175
Cricklewood 60
Crosby, Gordon 53
Czechoslovakia 166

D
Darracq 151
Day, B. I. 161
De Dion 40, 185
De Havilland 150
Delahaye 99
Delage 45, 88, 99, 170
Dixon 187
Domboy, Noel 165, 166
Don, Kaye 151
Dubonnet 94
Dunlop 67

E
Earls Court Show 134
Edgware 150
Edward VIII 93
Elliott, A. G. 162
Essex 103, 157
Evernden, H. Ivan F. 161

F
Fabroil 40
Falcon 160
Farina 113
Farnborough 169
Ferrari 131, 166
Fiat 165
Fisher & Ludlow 94
Fitzwilliam, The Earl 161
Flying Spur 125, 134
Foleshill 171, 172
Ford 171
Fottinger 169
Franay 113

Frazer-Nash BMW 155
Freestone & Webb 110, 124
Fulwell Engineering 169

G
Galibier 80
Gangloff 78
General Motors 93, 102, 111, 170, 171, 175
Germany 99, 166
Girling 43, 106, 169
Golden Arrow 169
Good, Alan 162
Gordini 165
Graham-Paige 173
Grinham, Edward 153
Grylls, Harry 171, 172, 176
Gurney Nutting 80
Guy Motors 169
Guy, Sydney 42

H
Halford, Major F. 150
Hassan, Walter 155
Harper, Ross 76
Harvey-Bailey, R. W. 160, 162
Havana 187
Heynes, W. M. 153, 155, 167, 168, 172
Hillman 150, 153, 168, 169
Hives, The Lord 150, 157, 158, 160, 162, 172
Hockley Heath 150
Hooper & Co 107, 123, 134, 175
Horch 88
Horton, O. D. 169
Hotchkiss 99
Houdaille 67
Humber 43, 93, 103, 107, 150, 153, 167, 168, 169, 171, 186
Humphrey-Sandberg 169
Hyatt bearings 183

I
I.A.E. 40, 161, 163, 168, 169
Invicta 72
Irving, Capt. J. S. 169
Isle-of-Man T.T. 11

J
Jaguar 45
Jelf Medalist 160
Jensen 111
Johnson, Claud 149, 161
Jowett 161

K

Kenilworth 89
Kestrel 72, 158
King's College London 160, 161
Kings Norton 176
Kingston-on-Thames 160
Klosternenburg 166
Knight, Charles Y. 45, 55, 57, 188
Koprivnice 166
Kortz, Felix 164

L

Lagonda 72, 75, 93, 161, 162
Lago-Talbot 88, 99
Lancefield 69
Lanchester 15/18 18, 51
Lanchester, Dr Fred 20, 25, 51, 52, 163
Lanchester Engine Co. 20
Lanchester, Frank 20
Lanchester, George 20, 163
Lanchester Laboratories 51
La Salle 90
Le Canadel 160
Leamington Spa 89, 167
Ledwinka, Hans 166
Leyland Eight 7
Lincoln 89, 99
Lisle 40, 42
Liverpool University 168
Lockheed 106, 178
London University 160
Longbridge 94, 99
Longmore 189
Lorraine Dietrich 181
Luton 171
Luvax 55
Lyons, Sir William 153, 172

M

M.G. 172
M.O.T. 102
Magdalen Bridge 102
Manley, H. B. 43
Marles 86, 150, 177, 179
Martin, Percy 50, 51, 185, 188
Maybach 88
Mazak 93
Meadows, Henry 169
Mercedes-Benz 99, 181
Merlin 158, 160
Metalastic 127
Michelin, Pierre 176

Miller 164
Minerva 45, 70
Molsheim 75, 78, 164, 165, 166
Moorfield Works 42
Moray, The Earl 69
Morris 19, 170
Moult, Dr E. S. 150
Mulliner, Arthur 55
Mulliner, H. J. 25, 29, 107, 113, 124, 134, 140
Munich 166

N

Napier Company 27
Napier Lion 158
National Motor Museum 169
Nesseldorfer Company 166
Newton 55
Northampton 55
Nuffield Group 172

O

Oldsmobile 102
Olley, Maurice 171
Olympia Show 51, 54, 93, 153, 170, 173
Oxford 102

P

P.100 124
Packard 90, 99, 102
Paris 64, 70, 151, 165, 166, 170, 176, 181
Park Ward 64, 96, 107, 113, 123, 129, 131, 140, 155, 172
Parkside 111
Paulin, Georges 160
Peregrine 72, 160, 173
Perrot 25, 31, 67
Peterborough 158
Pichetto, Antonio 165
Pomeroy, Lawrence 50, 51, 54, 57
Porsche, Ferdinand 166
Pratt, Jimmy 43
Pressed Steel Co. 107
Provence 160
Pugh family 20

R

R.A.E. 169
R.Ae.S. 162
Radford 45, 51, 53, 57, 169
Railton, Reid 53
Repton 162
Rhine 88
Ricardo, Sir Harry 103

Riley 65
Robotham, W. A. 162, 163, 171, 172, 176
Rolls, Hon. C. S. 11
Rootes Group 93, 169
Rover 65, 80, 103, 171
Rowledge, Arthur John 158
Royal Collection 57
Royal Household 48
Royce Cranes 50
Royce, Sir Henry, Bt 11, 72, 149, 150, 157, 158, 160, 161, 163, 173, 182
Rudge-Whitworth 20, 26, 73

S
S.S. Cars 80, 168, 171, 172
S.U. 103, 127, 150
Sapphire, Armstrong-Siddeley 111
Schneider Trophy Race 182
Scintilla 68, 76
Sedburgh 168
Segrave, Sir Henry 169
Sentinel 163
Serck 78
Sheffield-Simplex 161
Shrewsbury 163
Sidgreaves, Sir Arthur 72, 158, 162, 174
Silentbloc 55, 65, 102
Skefko bearings 182
Sleator, Walter 175
Small Heath 150
Smith, Jack 70
Smith-Clarke, Capt. 76
Solex 68, 182
Solihull 76
Sparkbrook 20, 25, 26
Staines 169
Standard Motor Co. 153, 163, 168
St Denis 176
Stevens, H. C. 169
Steyr 166
Strasbourg 165
Stromberg 86, 103
Stuttgart 99
Sussex 157

T
Tait, Dr Noel 168, 169
Tatra 166
Taub, Alex 171
Thompson Group 117
Thompson & Taylor 53
Thorneycroft 169
Tiffins 160
Timken Magazine 187
Tresilian, Stuart 162, 178, 179, 182
Triumph 80, 169
Turin 165
Turner, Edward 169

U
U.S.A. 34, 57, 58, 170
Umberslade Park 150

V
Van Den Plas 80, 95
Van Vooren 70
Vauxhall 51, 54, 161, 171
Vauxhall Motors 43, 51, 54
Vickers 163
Vienna 166

W
Wall Street 58
Wallasey High School 168
Warwick 55, 167, 168
West Wittering 160, 161, 162
Weybridge 53
Weymann 29, 55
Wilks brothers 171
Willesden 134
Wilmot Breeden 80
Wilson, Capt. 53
Wilson gearbox 53, 169
Wishart, Jack 167
Wolseley 18, 80, 158, 163, 172
Wolverhampton 28, 30, 40, 117, 151, 167, 170
Worcester 155

Y
Young, James 80, 110, 125, 134